*Dictionary
of
Library and Educational
Technology*

Dictionary of Library and Educational Technology

2nd Edition, Revised and Expanded

Kenyon C. Rosenberg
Associate Director
Bibliographic and Document Services
National Technical Information Service

with the assistance of
Paul T. Feinstein
Chief
Bibliographic Services Division
National Technical Information Service

1983
LIBRARIES UNLIMITED, INC.
Littleton, Colorado

Copyright © 1976 Kenyon C. Rosenberg and John S. Doskey
Copyright © 1983 Kenyon C. Rosenberg
All Rights Reserved
Printed in the United States of America

No part of this publication may be reproduced, stored in a retrieval system, or transmitted, in any form or by any means, electronic, mechanical, photocopying, recording, or otherwise, without the prior written permission of the publisher.

The first edition was published under the title *Media Equipment: A Guide and Dictionary*, by Kenyon C. Rosenberg and John S. Doskey (1976).

LIBRARIES UNLIMITED, INC.
P.O. Box 263
Littleton, Colorado 80160-0263

Library of Congress Cataloging in Publication Data

Rosenberg, Kenyon C.
   Dictionary of library and educational technology.

   Rev. ed. of: Media equipment. 1976.
   Bibliography: p. 171
   1. Audio-visual equipment--Catalogs. 2. Audio-visual equipment--Dictionaries. I. Feinstein, Paul T.
II. Rosenberg, Kenyon C. Media equipment. III. Title.
TS2301.A7R66 1983   621.38'044   83-19641
ISBN 0-87287-396-X

Libraries Unlimited books are bound with Type II nonwoven material that meets and exceeds National Association of State Textbook Administrators' Type II nonwoven material specifications Class A through E.

To the most important women in my life
(in descending chronological order)
Esther, Jane, Lorna, Victoria, Dana, and Katherine
Rosenbergs, every one

Table of Contents

PREFACE TO FIRST EDITION................................ix
INTRODUCTION AND ACKNOWLEDGMENTS..................xi

CRITERIA FOR EQUIPMENT SELECTION...........................1
  Projectors..1
    Motion Picture Projectors...1
    Super 8mm Loop Projectors.......................................4
    Slide Projectors..4
    Filmstrip Projectors...7
    Overhead Projectors...8
    Opaque Projectors...9
  Sound Systems..9
    Phonographs...9
    Turntables, Tone Arms, and Phono Cartridges....................11
    Amplifiers...14
    Tuners and Receivers...14
    Loudspeakers...15
    Tape Recorders...16
  Videotape Systems...20
  Reproduction Equipment..21
    Reprographics..21
    Microform Equipment..22
  Computers...24
    Hardware...24
    Software...25
    Optional Equipment and Applications............................26

DICTIONARY OF LIBRARY AND EDUCATIONAL TECHNOLOGY....31

BIBLIOGRAPHY..171

Preface to First Edition

Most persons involved professionally in the field of educational media must usually possess at least a rudimentary knowledge of principles and demonstrate skill in handling basic types of audiovisual equipment. The work of a somewhat smaller number of such persons requires more comprehensive familiarity with the technicalities of media equipment. As the list of media devices grows longer and more complex, both groups are likely to experience problems of communication and a need to be more capable and precise in their use of the technical vocabulary that accompanies them.

Because the field of instructional technology and educational media is now undergoing the expansion of scope and numbers that it is, these problems of communication are compounded. Nowhere is this more evident than in the processes of examining, testing, and purchasing items of media equipment required to support educational, information, and training programs of schools, colleges and universities, public libraries, industrial and business organizations, religious institutions, and myriad other agencies throughout the country. In all such units, and in the absence of a suitable consumer's guide for the purpose, media professionals continue to organize competitive demonstrations and tryouts of audiovisual equipment and to recommend for purchase those that match locally imposed technical criteria and financial constraints. In doing this, they must read (and understand) numbers of specifications, claims, and counter-claims; and they must obtain as much credible information as possible about comparative strengths and weaknesses of competing items and brands. In all such cases, exactness of meanings of technical terms encountered is an obvious essential.

In writing this book, Kenyon C. Rosenberg and John S. Doskey have presented us with a useful tool for improving our ability to carry out such tasks. Written in a manner that is both exact and understandable, and broad in its coverage, this book should be valued by all who work to achieve systematization and improvement of media equipment selection and evaluation and to develop better equipment maintenance programs. Media professionals will find it a valuable source of explanations and definitions of the technical data needed to prepare equipment bid requests, to read and compare technical specifications "on paper" before engaging in "hands-on" testing, to develop media hardware service contracts, or to carry out other necessary tasks.

Perhaps the book will be best used in conjunction with other helpful publications. Among these are the *Audio-Visual Equipment Directory* (National Audio-Visual Association, 3150 Spring St., Fairfax, VA. 22030); certain of the *EPIE*

Reports (Educational Products Information Exchange, Inc., 463 West St., New York City, 10014); *Educator's Purchasing Guide* (North American Publishing Co., 134 No. 13th St., Philadelphia, Pa. 10107); *Library Technology Reports*; and the media hardware reviews contained in *Audiovisual Instruction*.

The authors' credentials for producing this book are well worth citing. Rosenberg's long years of experience in the electronics and engineering information fields, as well as in library and information science, enabled him to serve regularly as column editor (Hardware Reviews) of *Previews* magazine. Doskey's experience in instructional technology and teaching has given him the background to understand and appreciate the practicing educator's need for the assistance that this book provides.

1976 James W. Brown

Introduction and Acknowledgments

The work in hand constitutes the second edition of *Media Equipment: A Guide and Dictionary*, by Kenyon C. Rosenberg and John S. Doskey (Littleton, Colo.: Libraries Unlimited, 1976). This edition has been completely revised and updated to include technical terms appropriate to the fields of reprography, micrographics, communications, and computers. The number of terms herein defined is virtually double that in the previous work.

Libraries and schools have, in the last fifteen years, been thrust (and sometimes dragged) into that unknown area—color it gray—of "technology." Many library and education practitioners have had to learn a goodly amount regarding the selection, use, and care and feeding of such diverse equipment as phonographs, tape recorders, projection devices, video tape players, copying machines, microform readers and printers, and computers.

Not only has the working professional not had available, in one vade mecum, descriptions of how these things worked and some comprehensible definitions of some of the most often encountered terms, but neither has the student had—written, insofar as possible, in non-technical terms—access to this information. It is this lack which the author hopes to rectify.

This work comprises three sections. The first is a set of descriptions and, where possible, groups of selection criteria specific to the type of equipment under consideration. No such list can be all inclusive; this one should be construed as a kind of selection guide for those who have little or no knowledge about the types of equipment which they may have to select or purchase.

The second, and larger section, is an alphabetically arranged (with abbreviations and acronyms at the beginning of each letter) dictionary of almost 800 terms, organizational names, specifications, etc., found in connection with the types of equipment mentioned above. Obviously, when an author does not set out to create a work that is comprehensive (especially a lexicographical work), readers may come across terms, etc., not to be found between the covers of that work. For that, please accept, in advance, the author's humblest apologies. The terms included here have been selected as thoughtfully and carefully as possible, and many of those omitted were found to be (A) admirably defined in the usual, general English language dictionaries; (B) so specific as to interest only the specialist; (C) not encountered often enough.

Following the dictionary is a selective bibliography, arranged by type of equipment. This should prove helpful to those who wish, or require, additional readings.

Readers are thanked in advance for their indulgence and patience regarding any errors (or obvious omissions) they may discover, and for any suggestions for future editions.

To my friend, John S. Doskey (my previous co-author, and eminent William Maclure scholar): thanks for the technical and moral support.

Last, to my wife, Jane: again, thanks for making everything better and easier.

Criteria for Equipment Selection

This section provides information about the evaluation of various types of media equipment. In considering the purchase of a particular piece of equipment, the considerations mentioned in the Criteria section should be kept in mind. For purposes of comparison, other information about media equipment may be found in publications of the Educational Products Information Exchange (EPIE) Institute and in the National Audio-Visual Association's (NAVA) *Audio-Visual Equipment Directory*.

PROJECTORS

Motion Picture Projectors

In almost all schools, libraries, and media centers in the United States, the first choice element with respect to **motion picture projectors*** is the size of the **film** to be used. These are essentially two: 16mm (the most popular) and Super 8mm. (There is a standard 8mm film, but it is fast becoming obsolete.) The size designation of each film relates to its width. Super 8, devised by Kodak in 1965-66, offers 50 percent larger **frame** size than the old 8mm, while utilizing the same width film. (The Super 8 frame is three-eighths the size of 16mm, while the regular 8mm frame is one-fourth the size of 16mm.) Both types of film are designed for projectors with **sprocket** wheel drives.

The general description that follows is appropriate to both 16mm and 8mm motion picture sound projectors. Both such units are made up of three related systems, each of which will be treated separately.

The Mechanical System

The function of the mechanical system of the motion picture projector is to move the film from one **reel** to another. To do this well, it must move it at a predetermined rate, which is expressed in **frames-per-second (fps)**, and with minimum inconsistency. The hallmarks of a good motion picture mechanical system are consistency of speed, quiet operation, and gentle film handling (i.e., does not damage sprocket holes or break the film). This last function can be tested simply by running new film through the machine, then checking for scratches and broken sprocket holes. Also, the projector should be able to handle old or damaged film without malfunctioning.

---

*Terms defined in the Dictionary within this book appear in boldface the first time they occur in a section.

2 / Criteria for Equipment Selection

Naturally, a mechanical system, which must accomplish many things at once, will create some **noise**. Therefore, a good test of this system is to operate the projector in a quiet room, with the sound turned off, while sitting close to the device as it runs. If it is objectionably loud under these conditions (the noise level should be forty-four **decibel**s or less, as measured with a simple sound-pressure-level meter), try turning up the sound and see whether, at a normal listening level, the sound sufficiently masks the noise produced by the mechanical system.

The mechanical system of a good projector will not damage the film. However, the threading process should be clear and simple enough to prevent misthreading and consequent film damage. Also, if the projector is designed for automatic threading, it is advantageous if it can also be manually threaded, and/or unthreaded before the film is completely finished.

Some good projectors do not have sprocketed (geared or toothed) wheels to drive the film, but instead use rubber wheels that grip the edges of the film and drive it frictionally. Unfortunately, many of these types of projectors do not drive the film with the consistency of speed that can be achieved by the sprocket system.

The Optical System

The components of the optical system of the projector are usually a light source or **projection lamp**, a lens (or lenses), a "framing" device, a fan to cool the light source (driven by the motor of the mechanical system), and a **shutter**, which "blacks out" the **screen** between frames. The function of this system is to direct a focused beam of light through the film as it moves by the film **gate**, and then to pass this **image** through the projection lens and project it upon the screen in a viewable size.

The critical aspects of the optical system are as follows: the light source should not overheat (and burn) the film; the shutter should operate properly and quietly; the lens should cast a sharp, clear image of appropriate size.

The lens is a key factor in achieving good motion picture projection. It should have at least a two-inch **focal length** (50mm is standard), and a **lens speed** of between f/1.3 and f/1.6. The lenses of most modern sound projectors should have a **resolving power** (clarity of the projected image) of between 50 and 80 lines/mm. Less than 50 lines/mm will result in a markedly fuzzy image. When selecting a projector, check the projected image not only in the center, but also around the edges and corners. Because of the necessary convexity of the front of the lens, uniform resolution is virtually impossible to achieve. Center resolution should average about 180 lines/mm, and high is 200 lines/mm or better. Corner resolution should be at least 50-80 lines/mm when the lens is focused, and 150 lines/mm is excellent.

Some projectors are equipped with, or have as an option, a **zoom lens**. This can be very useful for projecting an image of appropriate size at various distances from the screen. However, if the projector is to be used in one room, at a fixed distance from the screen, the zoom lens is not needed.

The quality of illumination (brightness) of the optical system may be measured with a **photometer** (light meter). The degree of brightness depends on the type of lens used and the light power of the bulb or lamp. The light output should be at least 416 **lumen**s or more. A coated lens usually presents a brighter image than an uncoated one. Very even brightness occurs when corners are at least two-thirds as bright as the center; one-half as bright is fair. With respect to the light source, a low-voltage tungsten-halogen lamp usually provides the longest life and maximum brilliance. Using a 250-watt tungsten-halogen lamp, you can expect a minimum of twenty-five hours of use on high output. The cooler the light source is kept, the less chance there is of damaging the film and, as a by-product, the longer the bulb will last.

The best type of fan for cooling the light source seems to be the type called the "squirrel cage." It moves considerably more air than the bladed type, and it is usually quieter. A quick and simple way to check the fan's efficiency is to run the projector for about ten minutes and then place one's hand near the ventilation grill. Air being expelled should be noticeably warmer than the ambient air in the room. This means that the bulb is losing heat effectively. A further test of the effectiveness of the cooling system is to check the temperature in the film gate area. It should be between 100 and 110 degrees Fahrenheit; 118 degrees or more is unacceptable.

If the projector has the capability of "freeze framing" (that is, **stop-motion**), some apparatus (usually a glass plate) should be interposed between the bulb and the lens to prevent the film from burning. Also, the bulb voltage should drop automatically. Try this function for about three or four minutes to check for any noticeable degradation of either the film or the image.

The Sound System

In older motion picture projectors, this is exclusively an **optical sound track** system. The sound is converted into optical impulses and laid down, as a "sound track," alongside the frames. In the projector, a light beam from the **exciter lamp** passes through the film sound track and is projected onto a **photoelectric cell** in the sound drum (on which the film is moving). The cell changes the varying light impulses to electrical impulses, which are then amplified, and these electrical impulses activate the projector's **loudspeaker** system.

A number of factors should be considered when evaluating the sound system of a motion picture projector. Is the speed of the film over the sound drum sufficiently constant? (The answer to this question depends on the consistency of speed with which the mechanical system runs.) Is the speaker large enough, and of sufficient quality, to reproduce the sound for the desired room size or audience? Is there provision for plugging in an extra speaker — or a **microphone**? Finally, is the **power output** of the **amplifier** adequate for the room or auditorium size in which the projector will be used?

Although most 16mm sound motion picture projectors use optically recorded sound, virtually all 8mm types use **magnetic recording** in the form of a **stripe** (which is exactly the same system used in all tape recording). Many newer 16mm sound projectors are equipped to handle both types of sound recording, while some offer only the magnetic type. Not all projectors that can use magnetically recorded sound, however, can also record. Some have playback devices only, while the more flexible ones allow for recording also. These are more flexible in that one can change the sound track at one's pleasure. The most sophisticated of the projectors that use magnetic recording not only permit simple recording from a microphone, but also allow for "line and mike mixing." This means that the microphone and another sound source (**phonograph, tape recorder**, etc.) can be mixed together in a controlled **volume** blend to produce both sources on a film. The magnetic recording system, generally, when used properly, produces better and cleaner sound. Its major disadvantage lies in the possibility that one may inadvertently erase the existing **program**.

A valuable source of test films for 16mm sound motion picture projectors is The Society of Motion Picture and Television Engineers (862 Scarsdale Ave., New York, N.Y. 10583). These films may be used to test comparative performance of 16mm projectors with respect to such factors as amplifier and speaker response to low and high frequencies, steadiness of screen image, relative distribution of light at various points on the screen, intelligibility of dialogue, reproduction of music without **wow, flutter**, etc. It should be emphasized that all tests of competitive projectors must be made under the same conditions and with the same films.

Super 8mm Loop Projectors

The 1960s saw the development of the Super 8mm continuous-loop **film cartridge**. In this type of system, one simply inserts the cartridge into a slot in the projector and the machine is turned on. No threading or rewinding is necessary. Some such projectors are even prefocused and, when the **film** has been shown, turn themselves off, leaving the cartridge rewound for the next showing. Other special characteristics that have been incorporated by some manufacturers are: built-in projection **screen**s, **zoom lens**es, and recording capability.

Many of the criteria used to evaluate 16mm **motion picture projectors** are also applicable to the Super 8mm loop projectors. However, special problems are sometimes encountered in loop projectors, such as difficulty in retrieving either broken film or film pulled off the cartridge hub, bulb changing, and film jamming. The only way to test for these problems is to run the loop projector for a number of hours (at least fifteen) using different films—some short and some long—to see how well it handles these varying lengths. Another point to consider is that, as yet, there are fewer films available in the cartridge format.

Slide Projectors

In general use today is the "miniature" or 2x2-inch **slide** in which is mounted a single **frame** of transparent 35mm **film**. These are termed "miniature" because, earlier in this century, the "regular" lantern slide had dimensions of 3¼x4-inches. Since the advent of the popular 35mm camera, the 2x2-inch slide has virtually supplanted the larger size.

A **slide projector** generally consists of two principal systems: a mechanical system for moving the slides and an optical system for projecting the slide **image**. The purpose of the mechanical system is to move the slide from its carrier or container (the latter may be either a circular or a cube-like tray) to a position in the optical system from where it may be projected. There are two basic systems for accomplishing this. The older one is the "mechanical feed" type, in which the slide is manually pushed or pulled into the optical system by means of the carrier. In the newer system, the so-called "gravity feed" (developed by Kodak in its Carousel projectors), the slide is allowed to drop (hence "gravity") into the optical system from its tray; when the viewing time has elapsed, it is mechanically pushed back into the tray. Both systems have good and bad points. The gravity system reduces physical handling by half, and slides seem to suffer less wear. The major problem with gravity systems is that if the projector is tilted, sometimes the slide does not drop.

A number of standards must be considered when evaluating the mechanical aspects of a slide projector; these standards are applicable to both manual and automatic machines. Will the slide carrier or container accept both paper-mounted and glass-covered slides? In the case of automatic projectors, does the container accept the number and types of slides you desire? Are the containers reasonably priced? May they also serve as storage units? Is there a locking device to prevent accidental spilling of the slides? Does the projector have an adequate framing device? Does it have an adequate elevation mechanism to adjust the image up and down on the **screen**?

Many of the newer gravity feed slide projectors are automatic in that they can be set to show slides in a prearranged time sequence. Also, one of the most useful options available with currently produced slide projectors is a remote control unit which provides for both forward and reverse motion, plus adjustment of focus. This allows the projectionist some freedom of movement or position, and it means that he can hear audience questions that he could not hear if he were sitting next to a projector with a noisy fan.

The slide projector and the **motion picture projector** have an optical system whose purpose is the same, i.e., to produce a well-defined image on the screen. It is apparent that the optical system of the slide projector has characteristics similar to those of the motion picture projector. However, since the slide produces a still picture that the audience views in detail and for a greater duration, the lens of the slide projector should have greater **resolving power** (60-100 lines/mm) than that of a film projector. Some manufacturers offer a selection of projector lenses of varying **focal length**s and **lens speed**s (f/ratings) to allow the projector to be used with different sized audiences, rooms, etc. Since the purchase price of these extra lenses may be prohibitive, a **zoom lens** is a very useful and important feature to consider when buying a slide projector.

In evaluating the optical system of a slide projector, you should use slides with small details for a sharp focus test; if possible, they should be projected onto a large screen. The image should be clear at the center and edges, and there should be no chromatic **aberration**. Illumination (brightness) readings may be taken with a **photometer** or, if no photometer is available, the evenness of the spread of light and the sharpness of focus may be judged visually.

There are other important factors to evaluate with respect to the light system of a slide projector. Does it accept lamps of sufficiently high or low wattage for the purposes intended? Tungsten lamps are less expensive, but they have a shorter life expectancy and tend to fade. Lower wattage (cooler) quartz lamps provide comparable brilliance with a system of reflectors and mirrors. Although they are more expensive, they will last longer and will not fade. Are there separate switches for lamp and fan, so that the fan can be used to cool the lamp after it has been turned off? This provision will increase bulb life. The effectiveness of the cooling fan can be tested in the same way as that of the film projector fan—by simply holding your hand close to the vent.

One of the main problems pertaining to slide projection is keeping the slides in focus once they are displayed. Unless they are mounted between glass plates, they will "pop" when placed in the optical system: heat from the bulb causes the slide to expand within its mount, forcing it to bulge at the center, and the projected image will then go slightly out of focus. Some projectors ameliorate this phenomenon somewhat by prewarming the three or four slides closest to the optical system so that when they are displayed, they will be, for the most part, in focus. An option on certain of the newer projectors is an automatic focusing device, which resolves this problem through the use of additional optical equipment.

In recent years "slide-tape" presentations have become increasingly popular. These are achieved by coupling a slide projector with a **tape recorder** or **audio cassette** player; slides in the projector are advanced automatically by an inaudible pulse on one of the **stereophonic** channels, while the second **channel** of the tape presents the **audio** message, narration, or music. There are a number of excellent self-contained machines now available on the market, in addition to individual synchronizers that can be interconnected with stereo tape and slide projector units (also, **filmstrip** projectors). Among the least expensive of these synchronizers are those available from the Edmund Scientific Co. (624 Edscorp Building, Barrington, New Jersey 08007), whose free catalog, by the way, is a browser's delight. One such unit can be used with either of those types of Kodak Carousel or Airequipt slide projectors which are capable of being operated by remote control. This Edmund unit is available with the essential connecting cable for the projector.

When considering the purchase of slide-tape equipment, one should decide whether an integrated unit or separate components would be more useful. But there are other factors to keep in mind when evaluating such equipment. Is the pulse-**program**ming easy to use and easy to re-program? Are the pulse frequencies compatible with other sub-systems that might be incorporated into the network? How reliable is the pulsing system? (In testing reliability, determine whether **signal**s are ever missed by the projector or see if the system occasionally double-trips the slides.) If the program does go out of **synchronization**, how easily can it be locked back in during the presentation?

Currently, a number of manufacturers do market 2x2-inch slide projectors which are unique in that they eliminate the need for a tape recorder or cassette player to provide sound. The slide (including, usually, its cardboard mount) is mounted in variously designed borders. These borders are possessed of magnetic striping, etc. (depending on the manufacturer's system), upon which can be recorded (usually through a recorder/player built into the projector) a brief message varying in length from ten to thirty seconds. At the end of the message, a pulse can be added so that, in actual use, the audience views the projected image and simultaneously hears the recorded message, at the conclusion of which the next slide is automatically advanced. Once more, the projection portion of the device should be equal to regular projectors, while the recorder/player — since, more often than not, it will be used only for spoken rather than musical material — need not have the **frequency response** capabilities of good, separate audio equipment.

A number of combination filmstrip and 2x2-inch slide projectors are available that permit switching from slides to filmstrips or vice versa. In some of the older model slide projectors (usually manual), the slide carrier is removed and the filmstrip is fed into the optical system from a spool or other device mounted on top of the projector. Quite a few filmstrip projectors have, as an accessory, a slide carrier that converts the machine into a slide projector. Some do not have such an accessory, however, so the slide has to be placed in the optical system manually, usually with a certain amount of fumbling. Thus, it is important to consider whether the projector design facilitates quick, foolproof change from filmstrip to 2x2-inch slide projection or the reverse. Secondly, do both the slide and the filmstrip carriers have pressure plates or other devices for holding the slide or filmstrip at the focal point? Finally, will the filmstrip carrier accept both single- and double-frame filmstrips?

Filmstrip Projectors

Filmstrips, until recently, were called **slide** films. Essentially, a filmstrip is a group of 35mm transparent **frame**s placed in sequence on one piece of **film**. (There are also 16mm and 8mm filmstrips, and they are increasing in use.) Filmstrips are frequently differentiated as being either "silent" or "sound." This is actually an empty distinction, since filmstrips do not have an **optical sound track** or magnetic sound **stripe** on the film itself. Sound filmstrips are intended to be used with a recorded **program**, usually on a **disc** (record) or an **audio cassette** tape. Silent filmstrips are not used with a recording and often have an explanatory or narrative caption on each frame.

The projectors for "silent" filmstrips vary greatly in design but can be divided into two basic groups. The first is what may be termed "semi-automatic" framing projectors. Initially, the filmstrip is inserted into the machine (or carrier) and the focus frame, title, or picture is brought into view for focusing. If the **image** is split between two frames or is not precisely centered in the **aperture**, it may be adjusted with the framing control (see **framing**). This is usually a lever or some other mechanism that is pulled, depressed, or otherwise manually operated. Thus, each subsequent picture or image should be precisely framed as it appears on the **screen**. Although a properly acting semi-automatic framing device is desirable, many of these tend to go out of frame or "half-frame" after a few advances. This means that the operator must repeatedly reframe the film.

A second type of filmstrip projector is the continuous "manual framing" type. With this device, the operator usually rotates a knob to advance the film, and this manual rotation determines proper framing for each picture as it comes into view. For some very inexpensive (read "cheap") projectors, the operator, instead of advancing the film by a rotating, knob-controlled mechanism, pulls it through the projector with his fingers. This type of projector is not recommended because finger marks and dirt may get on the film and because, under certain circumstances, the **sprocket** holes may be damaged.

Although there are two basic types of filmstrip projectors, they use a variety of internal devices that actually move the film. Some have sprocket wheels, which mesh with sprocket holes in the film and act as a driving system. Others use a **pawl-sprocket** mechanism to advance the film. Perhaps one of the best means for moving the filmstrip is found in those projectors which use rubber rollers to grasp the edges of the film gently. Most "semi-automatic" framing projectors rely on either the pawl mechanism, sprocketed wheels, or a combination of the two. But these methods can damage improperly inserted film. Thus, it is apparent that projectors that have manual framing rubber-roller type drives offer the best potential for gently treating what sometimes may be an expensive piece of **software**.

Currently, manufacturers offer two methods of presenting sound filmstrips. One is the coupled disc (record) player and filmstrip projector; the other is the coupled audio cassette recorder/player and filmstrip projector. Both types provide for an appropriate narration for each frame and, in many instances—equipment permitting—for advancing frames in automatic **synchronization** with the recording. The latter is accomplished with the aid of a pulsed **signal** on the disc or cassette **tape**, which may be either audible or inaudible. At the present time, virtually all professional programs on records are provided with dual recordings: automatic (inaudible pulses) on one side; manual (audible beeps) on the other.

Sound filmstrip projectors should have essentially the same kinds of desirable characteristics they would have if they were not coupled and were to be purchased as separate units. The only exception to this is that since the **audio** units will rarely, if ever, be used simply as music playback systems, their **frequency response**s need be only in the one hundred **hertz** to ten thousand hertz plus or minus four **decibel**s range. What does need testing in such devices is their ability to move the filmstrip properly so that the film is not damaged and to maintain the framing so that partial framing does not occur. If the projector part of these machines can be used as a simple, manual filmstrip projector when needed, this is an additional advantage. Whether one decides to purchase a silent or sound filmstrip projector, or one with semi-automatic or manual framing, the criteria and tests for evaluating the optical and light systems of film and slide projectors are applicable.

Overhead Projectors

The overhead **transparency** projector is a quite simple and effective visual teaching device that is easy to use and maintain. It projects an **image** on a **screen** by passing light through a translucent material (usually transparent acetate) which is placed on the projector stage or platform. This stage should be able to accept all standard sizes of masks and transparency areas up to a maximum of 10x10-inches. The light source is located under the stage, and the light, after passing through the transparency, is collected by the head-assembly above, then reflected through a lens to concentrate the light rays on the screen. The projected image should generally be bright enough for viewing in a lighted room.

Focusing of the **overhead projector** is accomplished by moving the head-assembly up or down and by changing the position of the reflector. It is important that the focusing system permit positive stopping at desired points without "drift," which may cause the screen image to slip out of focus. Also one should determine whether the projection and condenser lens system produces a sharp, flat-focus image and an even light distribution over the entire screen area. Again, as with other types of projectors, this judgment can be made visually or with the aid of a **photometer**.

Adequate illumination requires a light source of at least six hundred watts with a quartz lamp or one thousand watts with an incandescent type. Much of the stated wattage may be dissipated in the form of heat. Thus, it is important to note that the **lumen** output (measured by a photometer) of a lower wattage quartz-iodine lamp (which has less heat dissipation) may be greater than that of a higher wattage incandescent **projection lamp**. However, one should be very cautious when moving an overhead projector (usually the portable type) that has a quartz-iodine lamp. If the lamp has not been sufficiently cooled, the slightest jar may cause the filament to break. Whether an overhead projector uses a quartz or an incandescent lamp, it should have an adequate and quiet blower (fan) system with a separate switch (or switch position) to allow cooling the unit without running the projector lamp at the same time.

Overhead projectors come in a variety of sizes to serve different instructional situations. The larger, heavier types are designed for large-group instruction and are more or less permanently situated, or placed on an A-V cart. Some intermediate and less expensive types are easily carried and can be used for instructing either small or large groups. The 3M Company produces such a machine, which is adequate for small groups or the conventional classroom. Some of these machines come equipped with a cellophane roll attachment which allows the operator continuously to write notes, formulate outlines, etc.

All types of overhead projectors are particularly subject to the **keystone effect** (i.e., distortion of the projected image). Keystoning can be eliminated by tilting the screen forward; devices are available that can be attached to wall-mounted screens to serve this purpose.

Opaque Projectors

The **opaque projector** uses the principle of light reflection to create an **image** of flat, printed, or drawn pictures or other materials as well as some three-dimensional objects. Light from a high-intensity **projection lamp** is reflected from the object into a reversing mirror, which then passes it through a lens to the **screen** to create the image. An advantage of this system is that the user need not convert the objects to be projected into any other format—it is a one-step process. The great disadvantage is that since projection is by reflected light, which is much less efficient than transmitted light (the kind used for **slide**s, **film**s, etc.), successful projection with opaque projectors can be accomplished only in almost total darkness. Also, printed or pictorial material must be kept simple, so that every element, when projected, will be big enough for each viewer to see clearly. Opaque projectors are best used with small groups.

Materials to use with an opaque projector are usually placed on a platen, or platform, which can be raised or lowered to accept them. It is important that there be at least a 10x10-inch opening for reflection of flat pictures, etc. There should also be a heat-resistant glass plate, which will protect projected materials from the heat and hold them flat. Although the opaque projector has its limitations, the same focusing and illumination tests that were suggested for other projectors can be used. In most machines focusing is accomplished by rack-and-pinion gears. There should be a positive stop to prevent the lens from being popped out onto the floor. Some machines are equipped with a built-in light arrow or pointer device, which can be a useful option.

SOUND SYSTEMS

Phonographs

The **phonograph** is one of the most common of all media devices. It is designed to play back the sound encoded on the surface of a phonograph **disc** (or recording). The phonograph is composed of a number of sub-systems or units, of which the most important are the **turntable**, the **tone arm**, a **phono cartridge**, an **amplifier**, and a **loudspeaker**. In a **stereophonic** phonograph, there are two amplifiers and two loudspeakers. In this section, each of the sub-systems of the phonograph is treated separately. However, there are some specific aspects of the phonograph as a whole that ought to be remembered.

First, a phonograph is no better than any one of its sub-systems. If everything is of fine quality except, say, the phono cartridge, no disc will sound quite as good as it might if a better cartridge were employed. Much like a chain with a weak link, the phonograph ("gramophone" in Great Britain) cannot be a strong system unless all its parts are strong.

Second, phonographs designed for institutional use should be portable, rugged, and versatile enough to meet the institution's needs. Here portable is taken to mean weighing not more than twenty pounds. Rugged means that the device should be housed in a container that is not easily dented or marred, that the tone arm should be securable so that it does not flop about (and do injury to the phono cartridge when the machine is being carried), that the mechanical and electrical/electronic portions of the device should not be subject to being damaged if the unit is banged or bumped (as, say, on a cart that bumps heavily over elevator thresholds), and that the device should be able to play a disc when the disc is (moderately) warped or when the device is positioned slightly (up to twenty degrees) off the horizontal.

Versatility means, first, that the phonograph can play whatever recordings are necessary, regardless of their speeds (of which there are four: 78, 45, 33⅓ and 16 rpm). This does not necessarily mean that the unit needs to be able to operate at all of those speeds, but it must be able to accommodate recordings at speeds the institution deems necessary. Also, the device should be able to handle both **monaural** and **stereophonic** recordings, if required. A key point here is that the more options (e.g., playing speeds) a device has, the more complex (and subject to breakdown) it becomes.

Versatility may also mean the incorporation of such refinements as **cueing**, the ability to vary the speed of the turntable slightly from the selected speed, the inclusion of some **strobe** mechanism (which allows the user to check the speed accuracy), the provision for adding an extra loudspeaker by some simple means and, possibly, **automatic shutoff**. There should also be separate **tone controls** (i.e., **bass** and **treble**). In order to be useful in the classroom (or in other rooms of that size), the phonograph amplifier should have a **power output** of at least five watts per **channel**.

Most classroom phonographs are adequate only for playing recordings of spoken materials. Recordings of music played on most of these units will sound poor—that is, tinny and weak. If it is the institution's aim to develop the listener's appreciation of certain types of music, or to display specific aspects or nuances of music, these aims will not be met by the ordinary portable, institutional phonograph. This is especially true nowadays, when a large number of people (children and adults alike) have access to far better sound equipment than is usually available in institutions. It is for this reason, and for their greatly increased flexibility, that we recommend modular phono systems, which are made up of the separate sub-systems mentioned at the beginning of this chapter.

A cautionary word here: automatic record changers (i.e., the devices that can automatically play through a number of discs stacked on the spindle of the device) should definitely be avoided. The reasons for this are given in the following section, "Turntables, Tone Arms and Phono Cartridges."

Still, if a phonograph must be purchased, a few minimum specifications follow:

1. Power output of at least five watts per channel.
2. Stereophonic channel separation of at least twenty **decibel**s.
3. Total **wow** and **flutter** not exceeding 2 percent.
4. **Frequency response** of at least one hundred **hertz** to twelve thousand hertz plus or minus three decibels.
5. **Tracking error** not greater than three degrees.

Turntables, Tone Arms, and Phono Cartridges

The **turntable** is composed of the **turntable platter**, the motor to drive the platter, and usually the **tone arm**. The **phonograph** recording rests on the turntable platter, which is rotated by the motor. The tone arm, which is usually mounted to the right of the platter, contains the **phono cartridge**, held in a **head shell** (this should be detachable). The phono cartridge contains the **stylus** which, when placed in contact with the surface of the recording, follows the groove and is made to vibrate by the undulations in the groove. The phono cartridge converts these vibrations from mechanical energy into electrical impulses.

The turntable should offer a variety of rotational speeds (selected from 78, 45, 33⅓ and 16 rpm) adequate to the user's needs. Usually, only 45 and 33⅓ rpm are needed. The 78 rpm recording is a thing of the past, and the 16 rpm **disc** is almost never encountered except in the Library of Congress's "talking books" recordings. There are three types of drive systems for turntables. The least desirable (and most common) is the **rim drive**. A considerably better method is the **belt drive** type. Most expensive (but best of all) is the **direct drive** system.

One of the most useful options for a turntable is a speed adjustment control. This is particularly handy in situations where the turntable may be used for recordings that were made at a slightly erroneous speed, or where it is to be used by students for play- or sing-along recordings. Many such recordings are available; in order to be used properly, the note A given at the beginning of the disc must be tuned to 440 **hertz**. This, of course, is accomplished by adjusting the speed so that the A on the disc and that of a tuning fork (or a properly tuned musical instrument) coincide. In order for the speed adjustment to be truly useful, it should offer at least plus or minus 4 percent change. It is also handy if the turntable has a built-in **strobe** disc in order to allow the user to determine if the operating speeds are correct. The amount of speed error tolerable in a turntable is about 2 percent. The strobe disc can be used with a small neon lamp (of the circuit tester variety) to determine speed accuracy.

Basically, two types of motors are used in turntables. One is the **hysteresis** (or hysteresis synchronous) kind; the other, which uses salient poles, is usually referred to as an induction motor. Induction motors ordinarily come as two-pole or four-pole motors. Only the least expensive turntables use the two-pole motor. Some better turntables still use the four-pole motor, but the hysteresis (or hysteresis synchronous) motor offers a more constant speed than the other type and is definitely to be preferred.

The turntable should have a switch to turn the motor on and off, instead of the kind of switch that turns the device on when the tone arm is lifted from its rest. The tone arm should be of the locking type to secure it in place. The diameter of the platter should be great enough to support completely a twelve-inch disc.

The tone arm should be statically balanced. This means that instead of being spring-loaded it uses a counterweight to achieve proper **tracking force**. The spring-loaded type is considerably less expensive but commensurately less accurate. The tone arm should have a detachable head shell in order to facilitate changing the stylus or phono cartridge. The tone arm should have some method of mechanical **cueing** so that the operator can gently place the stylus on the required place on the disc.

The usual institutional turntable is supplied with a **piezoelectric phono cartridge**. This is used because, first, it is less expensive than the **magnetic cartridge**, and, second, it emits an electrical **signal** of greater **amplitude** (thus requiring less preamplification) than the magnetic type. The magnetic phono cartridge, however, offers substantial comparative benefits. It requires a lighter **tracking force** (thereby causing less wear of the stylus and disc). It is less subject to damage by rough handling or extremes of temperature or humidity. Further, the magnetic phono cartridge tends to offer a better **frequency response**.

If there is any intention of using these devices with the **quadraphonic** discs of the **CD-4** type, the cables that connect the cartridge to the **amplifier** should be the low-capacitance kind.

Lastly, it is wise to avoid purchasing the automatic changer type of turntable. First, the user usually wants to play only one of the two sides of the discs selected. Next, the changer-type turntable is far more likely to break down and to mishandle discs than is the manual turntable. Finally, the automatic changer turntable is often more expensive (because of the cost of the changer mechanism); it would be better to spend this additional amount on a manual turntable of good quality.

A good test of an integrated turntable (i.e., one that comes with a built-in tone arm) is to try playing a moderately warped disc (it's a good idea to keep one of these on hand just for this purpose). A well-designed turntable and tone arm should have little difficulty in properly playing a disc whose vertical warpage does not exceed one-half inch; it should be able to play the disc without audible thumps or skipping of grooves.

The purpose of the rubber mat covering the turntable platter is to prevent the cartridge from picking up any stray magnetism either from the motor or the platter itself. Therefore, another test of the turntable is to turn the motor on and hold the stylus about one-half inch above the platter. There should be no audible **hum** induced by this test. If, however, there is, it indicates that the cartridge is insufficiently shielded from stray magnetic fields. This is usually a fault of the turntable design and not necessarily a fault of the cartridge.

When testing a combination of turntable/tone arm/cartridge, be sure to play a recording with very loud musical passages at the end (inner portion) of the disc. A good combination of components will play such a passage cleanly, with little (if any) audible **distortion**, and without skipping grooves.

Important specification minimums for a basic turntable (without tone arm) are **wow** and **flutter**, which should not exceed .2 percent while unweighted **rumble** should be forty-five **decibels** or greater and weighted rumble should be fifty-five decibels or more. Flutter should not be greater than .1 percent. For a turntable equipped with a tone arm, the **resonance frequency** should be ten hertz or lower, or eighteen thousand hertz or higher, and **tracking error** should not exceed two or three degrees.

Minimum specifications for a magnetic phono cartridge should include a **channel separation** of twenty-five to thirty decibels, a **frequency response** of from forty hertz to twenty thousand hertz plus or minus three decibels, and the tracking force should be between .5 and 1.5 grams.

Phonograph Recordings

Since phonograph recordings, at present, constitute the largest **audio software** investment of most institutions, a few words concerning these common miracles are in order.

First, the **disc** is subject to warpage if improperly stored. Proper storage consists of housing the discs vertically with firm, but gentle, side pressure: that is, enough pressure to keep the discs erect and steady, but not so much that an individual disc cannot be removed with relative ease.

Next, discs are prone to deterioration due to small amounts of particulate matter in the grooves. This dirt is injurious to the stylus also, and causes premature wear. The dirt tends to adhere to the disc's surface because of two factors: static electricity, and gummy materials found in the air (such materials are found in abundance in expelled tobacco smoke). This means that recordings must be kept clean (as a matter of fact, even "factory sealed" recordings very often have dirty surfaces). The charge of static electricity acts as a magnet to hold the dirt on the disc's surface.

Getting rid of the static charge is rather simple. Take a strip of aluminum foil about 1½ inches wide and long enough to extend about 1½ or 2 inches past the edges of the foam or rubber turntable mat, when the strip is placed across the mat's diameter. Next, poke a hole (a sharpened pencil does a good job) through the strip where the spindle hole in the mat is. Bend the portions of the strip that extend beyond the mat under the mat, with the large portion of the strip on the top of the mat. When the mat is replaced on the turntable platter and a disc is placed on the mat, this strip grounds the disc and helps to prevent the creation of the static electrical charge, thus somewhat reducing the adherence of dirt to the surface of the recording.

What is more difficult is to remove from the disc the dirt and sludge that are already present. It has recently been proven that the best results are obtained when a disc is cleaned with a specifically designed disc cleaner moistened prior to each use. These products are useful for cleaning recordings after they have been circulated, and, before storage. All of these, and other such products, usually come with a liquid cleaning solution. It is imperative that not too much solution be used for each cleaning, since the solution can build up and necessitate even more stringent cleaning. We have found pure distilled water to be the most efficacious cleaning solution, and also the least expensive.

A phonograph recording that is circulated usually has a useful life span of from three to six circulations, because of careless handling of the disc and the inevitable resultant scratches. It is thus a matter of judgment as to how much care should be given those discs that do circulate. But most assuredly discs that do not circulate should be given all possible care in order to prolong both their life and that of the stylus.

Amplifiers

The **amplifier** portion of an **audio** system is its chief electronic component. The function of the amplifier is to take the minuscule electrical **signal** (usually in millivolts) and increase that signal several hundred or thousand times in order for it to be of sufficient size to activate the **loudspeaker**.

The amplifier usually contains a preamplification section, which enables it to accept the very small signal of a **magnetic cartridge,** and it usually has an **input selector**, which allows the operator to choose from a variety of **input**s (e.g., **tape recorder, tuner, turntable**), plus such other controls as **tone controls** and a **volume control**. The amplifiers usually encountered in institutions combine the functions of preamplification and are often termed "integrated amplifiers." Those that are not of this type (i.e., "basic amplifiers") have no controls and must be connected to a **preamplifier**.

The greater the number of input devices that can be operated through a given amplifier, the more flexible it is; and the more varied the controls, the more useful it is. Tone and volume controls are imperatives, but such other refinements as a **loudness compensation** circuit are helpful. In normal hearing there is an apparent loss of low **frequency** audibility as **amplitude** (i.e., loudness) decreases, that is, as the volume is made lower, there seems to be a proportional decrease in the **bass** one hears. This apparent loss occurs in human hearing even when there is no real loss of bass frequencies. This phenomenon was graphed by two researchers, and the resultant graphs are called the **Fletcher-Munson curves**. The loudness compensation circuit (usually activated by a switch) provides sufficient bass boost to accommodate the human ear at lower listening levels. Another handy circuit included in many amplifiers is one that allows the user to **monitor** a recording as it is being made, assuming that the tape recorder used is compatibly equipped.

A headphone **jack** on the front panel of an amplifier is a very useful **output**. Likewise, some means of switching off the power to the loudspeakers in those instances when only **headphone** listening is desired, is useful.

The power an amplifier supplies to the loudspeakers to which it is connected is termed its **power output**. For a system that uses **high-efficiency** loudspeakers, the power output should be at least five watts per **channel (continuous power)** in order to fill a modest-sized classroom with sound. It is best to have a minimum of twenty or twenty-five watts per channel, since this will allow for the use of **low-efficiency** loudspeakers and should be sufficient to fill a good-sized room with sound.

Other minimum specifications for an amplifier include a **frequency response** of from ten **hertz** to twenty-five thousand hertz plus or minus .5 **decibels**; **channel separation** of at least thirty decibels; **intermodulation distortion** of less than .5 percent; and a **signal-to-noise ratio** of sixty-five decibels or greater.

Tuners and Receivers

An **audio tuner** is an electronic device that must be coupled to an **amplifier**. The tuner allows the user to select broadcast radio **signal**s from the air, and when the tuner is connected to an amplifier (and the amplifier to **loudspeaker**s), one has, in effect, a radio. Most tuners are designed to receive both **amplitude modulation** (AM) and **frequency modulation** (FM) programs. Some, however, are designed to receive only the latter. Modern tuners are designed to receive the FM **band** and demodulate **multiplex** broadcasts into **stereophonic** programs.

A **receiver** is simply an integrated **amplifier** (i.e., an amplifier/**preamplifier** in one unit) that includes a tuner—all of this forming one unit.

Ordinarily, only in the finest (and most expensive) installations does one find a system that incorporates a separate basic amplifier, preamplifier, and tuner. It is much less expensive to purchase (and maintain) a receiver. The separate component system, though, does offer greater flexibility and allows for component substitution in the event of the failure of a given component. If one part of a receiver fails, all the other parts (since they are on one chassis) go with it to the repair shop.

In order to tune in stations more easily, it is helpful if the device includes a **centering meter** for FM reception, and a **signal strength meter** for both AM and FM programs.

The requirements for the amplifier section of a receiver are the same as those given in the preceding section. The minimum specifications for a tuner (and the tuner section of a receiver) include a **sensitivity** of 2.5 microvolts or less, selectivity of at least fifty **decibels**; **signal-to-noise ratio** of sixty-five decibels or greater, a **frequency response** of at least twenty **hertz** to fourteen thousand hertz plus or minus .5 decibels, stereophonic **channel separation** (for stereophonic devices) of at least twenty-five decibels, **intermodulation distortion** of .3 percent or less, and **harmonic distortion** of .3 percent or less.

Loudspeakers

The **loudspeaker** is a basic component of any **audio** system. In "ready made" systems (like the **phonograph** or radio), there is usually only one **piston speaker**. For better sound quality, however, the loudspeaker system is recommended. Such a system incorporates two, three, or more loudspeakers in a single **enclosure**. The enclosure should be relatively mar-resistant and sturdy. The front should be covered by an acoustically transparent grille cloth that will keep inquisitive fingers and dust out. If the loudspeaker must frequently be moved, it should be mounted on a cart; most loudspeakers are not designed for portability and may easily weigh over forty pounds.

There has been a great deal of technical discussion devoted to judging loudspeaker qualities. We believe that even if a loudspeaker meets the minimum specifications given below, it is still a very good idea to audition it in the environment in which it will be used, and in conjunction with the other components with which it will be used. Try playing through it recordings with which you are familiar. The best types of **program**s for testing a loudspeaker are those that offer organ, piano, and chamber music, and speech. This latter should sound natural and not boomy or overly resonant. Listening tests should last at least a half hour, preferably an hour or more. This will give the listener a chance to determine whether "listening fatigue" (a nervousness or desire to stop listening caused by unnatural sounding loudspeakers) sets in. Any loudspeaker that occasions listening fatigue should be shunned, regardless of its specifications.

Among the most important minimum loudspeaker specifications are:

1. A **frequency response** of at least from forty **hertz** to eighteen thousand hertz plus or minus four **decibels**.

2. A **resonance frequency** of twenty hertz or lower.
3. Power handling (i.e., the ability to deal with **amplifier power output**) of at least twenty to twenty-five watts greater than the **continuous power** output of the amplifier for a low to moderate power amplifier (in the ten to forty watts range).

Two other things to be aware of with regard to loudspeakers are **transient response** and minimum power handling. Transient response refers to the loudspeaker's capability of reacting to short, sharp attacks of noise (**transient**s); percussion instruments such as tympani and cymbals usually produce these transients. The loudspeaker should reproduce these brief sounds without continuing to vibrate once the impulse has stopped. Loudspeakers that continue to vibrate after the cessation of the impulse are said to "ring," or to "have an after image." Drum taps should be sharp and fade quickly, cymbals should clash and not sound sandpapery or otherwise distorted.

Minimum power handling refers to a basic difference between two types of loudspeaker systems, namely **high-efficiency** and **low-efficiency**. **Efficiency** is not a measure of quality; rather, it is an indication of the amount of electrical power (in watts) the loudspeaker system requires in order to produce sounds. High-efficiency systems require from three or four to ten watts of power; low-efficiency systems usually require from fifteen to twenty-five watts. Therefore, it is critical that the minimum power handling requirements be known in order to match the loudspeaker system to the amplifier. Again, it is best to audition the device rather than simply to rely on numbers.

Tape Recorders

Open-Reel

The **open-reel** type **tape recorder** is now a standard **audio** device in virtually all libraries and schools. In this type of device, a **reel** of **magnetic tape** is placed on the **supply reel** spindle, threaded past tape guides and **head**s and connected to the **take-up reel**. The **tape transport** section of the unit moves the one-quarter inch wide tape (at a variety of speeds, selected from 1⅞, 3¾, 7½, or 15 **inches-per-second**) past the all-important heads. In almost all currently made open-reel machines, these heads are of the **quarter track** type. If the device is a portable one, it will also include an **output amplifier** and **loudspeaker**(s), plus recording amplifier(s) and meter(s). If the unit does not contain loudspeakers and output amplifiers, the device is termed a **deck**. The machine will probably have both **line input** jacks and **jack**s for **microphone input**s. It may or may not have a built-in **mixer** to allow for the blending and balancing of **signal**s fed in from the various **input**s.

The heads are a critical portion of the machine. In less expensive devices there are but two heads: an **erase head** and one that acts as both a **record head** and a **playback head**. This kind of machine may allow one to **monitor** the recording being made, but only by **source monitor**ing. Better machines have three separate heads: erase, record, and playback. This permits **tape monitor**ing.

Tape speed is also a critical function. As a general rule, the faster the tape speed the better the quality of the recording. The most useful speeds are 3¾ inches-per-second (for speech and long musical selections, like opera, and where true **high fidelity** is not crucial) and 7½ inches-per-second (for high quality recordings for broadcast purposes, etc.).

Better tape recorders have a three-motor drive (or **direct drive**) system, not **belt drive**. The tape is moved by a **capstan** instead of by the reel drive method. The device will include **tape lifters** to remove the tape from intimate contact with the heads during the **fast forward** and **rewind modes** (thus preventing unnecessary head wear). There should be **VU meters** to allow the operator to see that the recording volume settings are appropriate, and a **digital counter** is necessary for indexing the tape so that specific portions may be found.

Machines that will be used with a variety of tape formulations should have their **bias** adjustment controls accessible in order that the bias current may be set appropriately for the specific tape used.

On portable open-reel machines, besides output **volume controls**, there should be separate **tone controls** for **bass** and **treble**.

Some machines offer such options as **bidirectional playback** and/or **bidirectional record**. These terms mean that the device can play back, or record, or both, on both sides of the tape without turning the tape over, that is, without reversing the reels. These functions are possible if there is an extra playback head (for reverse playback), an extra record head (for reverse record), plus an extra erase head. If the machine can offer both playback and record bidirectionally, it will have a total of six heads. Another handy pair of functions are **automatic reverse** (associated with bidirectional playback, at least, and often also with bidirectional record), which reverses the direction of the tape in either the playback or record mode, and **automatic rewind** which, once one side of the tape has been played through, automatically rewinds the tape onto the supply reel. The last option, which should be considered a requirement, is a **pause control**. Such a control stops the tape motion almost instantly in either the playback or the record mode, but keeps the machine in the appropriate mode. When the pause control is activated a second time, the tape motion begins virtually instantly—maintaining the same mode (playback or record) as when the tape was stopped. This allows for a simple form of **editing** in the record mode, or allows one to speak about what has just been heard in the playback mode.

Other special, optional (and usually desirable) features are **sound-on-sound** and **sound-with-sound** recording circuits. The former allows the recordist to add one **track** of recorded material to another, a process that allows a single instrumentalist to become a one-man-band, so to speak. Sound-with-sound allows one to hear a previously recorded track while recording on a second track. This method is often found in listening centers and language laboratories: the student hears a pre-recorded instructor, and he records his own responses on the same tape.

The last useful option is a **headphone** jack, which is used for monitoring purposes or when loudspeakers are precluded because of the environment.

Minimum specifications for an open-reel deck should include **channel separation** of twenty **decibels** or more; a **frequency response** (at 7½ inches-per-second) of at least thirty **hertz** to sixteen thousand hertz plus or minus two decibels; **harmonic distortion** not exceeding .3 percent **intermodulation distortion** no greater than .3 percent; a **signal-to-noise ratio** of sixty decibels or greater; and combined **wow** and **flutter** of .15 percent or less.

Before going to audio cassette devices, a word about the **Dolby noise reduction** system is in order. The Dolby system is found more and more often as an option on open-reel tape recorders, while it has been available in a large

18 / Criteria for Equipment Selection

number of cassette machines for some time. It is well known that all magnetic tape has an inherent **noise** factor called **tape hiss**. This can be obtrusive (and often really annoying), and it becomes greater as tape speed is decreased. In order to minimize this noise, Ray Dolby devised the circuitry that bears his name. Dolby noise reduction systems are available in two basic forms: as integral components, available separately, and as built-ins in specific tape recorders. The Dolby encoding and decoding of tapes renders tape hiss virtually inaudible. In good open-reel machines using a tape speed of 7½ inches-per-second (or faster), the Dolby system adds very little. But at speeds of 3¾ inches-per-second—and cassette machines operate at only 1⅞ inches-per-second—the Dolby noise reduction system is a must for high quality sound. The difference between a relatively slow tape that is not "Dolbyized" and one that is—when played back through a device incorporating Dolby circuits—is truly astounding. The Dolbyized tape presents its **program** from an almost totally noiseless background; quiet or silent portions of the tape are just that—quiet or silent. Therefore, for those many institutions that use their open-reel machines to record and play back music at 3¾ inches-per-second (in order to conserve tape), Dolby noise reduction is warranted. And for institutions that use audio cassette machines for recording and playing back music, Dolby is an absolute must.

Audio Cassette

The audio cassette tape recorder is increasingly becoming the recording playback device that most institutions depend on. Cassette **software** is more easily handled and stored than open-reel magnetic tape, and the **hardware** for such software is relatively simple to use.

The Philips Lamp Company, which holds the patents on the cassette itself (and allows others to manufacture the cassette under license), has so designed cassettes and their appropriate hardware that they, unlike some aspects of open-reel tape recorders, are fully compatible. A **stereophonic** cassette can be played on a **monaural** machine, and a monaural on a stereophonic. This is another advantage of this unique **audio** development.

The magnetic tape for these devices is housed in a small plastic box and permanently attached to two miniature **reel**s (actually, hubs). The tape is drawn from one of these reels to the other. In other words, the cassette is a tiny enclosed replica of a pair of open-reels. In the record and playback **mode**s, the tape always moves at 1⅞ inches-per-second. At the rear of the cassette are two small plastic tabs, one for each side of the tape. When either or both of these tabs are broken off, the corresponding side(s) cannot be re-recorded—thus preserving a pre-existing program.

Several types of audio cassette machines exist. There is the common, monaural portable which, in itself, is a complete record/playback unit. There are portable playback-only devices (good for circulating with software). Then there are portable stereophonic units and stereophonic **deck**s (we have yet to encounter a monaural deck, although they may exist).

The audio cassette device itself is simple. Virtually all such machines have but two heads: an erase head and one that combines the functions of the record head and the playback head. The controls for these units have been standardized. Besides the common ones (for record, playback, fast forward, and rewind), there is almost always a **pause control**. Also, almost all cassette machines now include a digital counter, and a few more decks have been introduced that have built-in line input and microphone input mixers. A very few exceedingly expensive decks also have three heads (erase, record, playback), permitting tape monitoring. Only one such deck we know of offers the ability to record sound-with-sound. More important for most users is that the device has meters to indicate recording **levels**.

The tape speed of the cassette was mentioned previously as being 1 7/8 inches-per-second. Another aspect of the tape itself is the fact that its width is slightly less than one-eighth inch, as opposed to open-reel tape, which is one-quarter-inch wide. Cassettes come in a variety of playing/recording times, among which are 20, 30, 60, 90, 120, and 180 minutes. These times indicate the total amount of playing/recording time for both sides (e.g., a 60-minute cassette offers 30 minutes per side). For most recordings, we do not recommend using longer times than 60 minutes (90, if absolutely necessary); the longer the tape (and the thinner it must be), the greater is the tendency for it to jam or otherwise malfunction in the machine. By the way, one way to help perclude cassette jamming is to slap the side of a cassette sharply against one's open palm just before using it. The slap helps free up the hubs and keep them moving easily when placed in the machine.

Because portable audio cassette machines come in such a variety of types, it would take more space than we have here to discuss all possible variations. Suffice it to say that the device should be carefully checked to determine whether it has all the necessary options, without having too many. Since the sound quality produced by these machines is usually severely limited by the vastly undersized **loudspeakers** almost always found in them, it is wise not to judge the total device too harshly. If the portable unit is intended primarily for use as a recorder, rather than for playback, we can suggest a rather simple test for it. Use the device to record the kind of **program** materials for which it will be used. Then play back the program on a good quality deck, connected to other good components (i.e., **amplifier** and loudspeakers). If the sound is adequate when tested in this fashion, the only further considerations are the warranty and available service. If, however, the portable unit will be used for playback, it can probably be improved rather simply. To do this one needs a small (6 or 8-inch diameter) **high efficiency** loudspeaker to which a light, two-conductor cable has been connected. The loudspeaker should be housed in a small, wooden **enclosure**. Attach an appropriately sized **plug** (either a mini-plug or a submini-plug) to the free end of the cable. Plug this into the earphone **jack** of the audio cassette machine. Chances are that the sound of the unit will be improved greatly. If one wishes to associate this larger loudspeaker with the tape recorder permanently, the loudspeaker can be glued to the underside of the machine. Then, if you add a small carrying handle, it becomes suitable even for classroom use. One last test for portable cassette units is to record and play back slow piano music, which will allow you to determine whether the amount of tape speed error is great enough to be a problem.

For audio cassette decks, the following are minimum specifications (besides having Dolby noise reduction circuits, as described in the last paragraph of the section on open-reel tape recorders):

1. Frequency response of from fifty hertz to thirteen thousand plus or minus two decibels (non-Dolby, using regular tape).
2. Channel separation of at least twenty-five decibels.

3. Signal-to-noise ratio of fifty-five decibels or more.
4. Wow and flutter not greater than .2 percent.

VIDEOTAPE SYSTEMS

The recording of visual and aural **program** material on **magnetic tape** has come to be known as **videotape** recording. Most videotape systems are composed of three separate systems. The first is the **videotape recorder**. This device consists of the **tape transport**, the **video head, audio record head** (which may double as the audio **playback head**), audio playback head, recording **level** controls, **meter**s to display recording level, and various playback controls.

The next part of the videotape system is the camera, which, of late, is of the vidicon type (see **vidicon camera tube**). Most cameras devised for use with videotape systems have a built-in **condenser microphone**.

The last part of the system is the **monitor**. Essentially, this is nothing more than a television set; however, it may or may not have a built-in **tuner** which would allow the monitor to receive (and, thereby, the **tape transport** to record) broadcast television programs.

Many videotape recorders offer two rather handy circuits as options. The first, **automatic level control**, allows the machine, when this circuit is activated, to set the audio volume level automatically as the sound being recorded varies. The other circuit is the **automatic gain control**. This circuit, when activated, automatically compensates for varying light conditions. When both of these features are present, all the operator needs to do is focus the camera and activate the recorder as desired. Further, most of these cameras now feature a **zoom lens** for even greater flexibility.

There are two basic approaches to magnetic **video** recording. The original method, videotape recording (commonly called VTR), is an **open-reel** system. The newer method of magnetic video recording termed **video cassette** recording, is abbreviated as VCR. Here the magnetic tape is housed in a large plastic container; because the tape in a VCR system does not progress from one **reel** to another within the cassette (actually it usually goes to a reel contained within the machine), the system is analagous to the audio **cassette** system in that the user need not handle the tape for threading, or any other purpose.

Except for the method of tape handling, the VTR and VCR function in exactly the same ways. Here again the analogy of open-reel audio tape machines and audio cassette machines is handy. Therefore, we shall continue to discuss these two recording methods as if they were one, generically terming them magnetic video recording. Both methods are available as black-and-white or color systems; color, naturally, is somewhat more expensive.

Magnetic video recorders should allow for camera, **microphone** (separate from the condenser microphone built into the camera), and **line input**s. The last-named **input** allows the user to record from another magnetic video recorder or from a television set.

Among some of the many options available for magnetic video recording systems are **stop action** (the ability to freeze a given **frame** for scrutiny); **slow-motion**; a **digital counter** for indexing the tape (this is essential); audio **dubbing** (allows the sound portion to be altered without changing the video portion); and editing (the removal of undesired sections and/or insertion of additional portions, both audio and video). All of these are desirable, but each will somewhat increase the cost of the system.

Considering the rather high cost of a magnetic video recording system, and also the comparative complexity of operation of this device, it is wise to live with a system for at least a few days before committing oneself to purchase.

Some of the more important minimum specifications for a magnetic video recorder (disregarding the camera and monitor, for which specifications will vary tremendously depending on factors like price, size, etc.) are:

1. **Wow** and **flutter** should not exceed .25 percent.
2. Audio **frequency response** should be at least from fifty **hertz** to fourteen thousand hertz plus or minus 2 **decibel**s.
3. Video frequency response from one hundred thousand hertz to ten million hertz plus or minus 3 decibels.

REPRODUCTION EQUIPMENT

Reprographics

During the five centuries which have followed Gutenberg's re-invention (from the earlier Chinese) of movable type, man has spent a good deal of his time, fortune, and ingenuity attempting to create a device which could copy both printed (first only those obtained from typeset then typewritten) and holographic materials. He has found it desirable, or necessary, to copy not only narrative materials but also graphs, charts, tables, works of art, etc. Until the invention of **electrostatic copying** in about 1960, the best a copyist could employ was some photostatic technique such as the Verifax system of Eastman-Kodak. Such techniques offered a true photographic copy well reproduced and of stable quality. However, as the author (when but a mere stripling of a librarian) can attest, making such copies was exceedingly time consuming (anywhere from two to five minutes per copy) and required working with chemical solutions whose aromas did not leave the copyist when that person repaired home (further, these solutions could stain clothes, fingers, etc.). Too, the quality of the copy depended more on art and luck than science. **Copying machine**s have come a long way since then.

In any event, the arrival of the very first Xerox machine made many a photostatic copyist return to his or her place of worship with a deepfelt song of thanksgiving. These devices are too common today to require a great deal of detail as to their numerous merits. Suffice it to state that they are available in two basic types: electrostatic copying and **thermographic copying**. The latter was the first available (in such devices as the Thermofax) and produced a short-lived copy, i.e., one which faded with exposure to light and with the general ravages of the passage of time. Electrostatic copying came with that first Xerox machine and provided copies on bond paper, which copies were virtually of archival quality, i.e., they did not tend to be noticably denigrated by light or the flight of time.

Both systems have their advantages and drawbacks, but the electrostatic technique has many more advantages and fewer drawbacks than does the thermographic. The electrostatic copy tends to last longer; it is made on bond paper (and, therefore, can be written upon with pen or pencil); since it is made on uniform sheets of paper (as opposed to the **web feed** in most thermographic systems), it is more easily handled by a **collator**; the process tends to be faster; the weight of the paper can usually be varied (up to light card stock). The only real disadvantage in electrostatic copying is the price of the **hardware**; it is rather more expensive than that of thermographic devices.

The thermographic device, through its web feed system, offers variable length copies and its hardware, as stated above, is less expensive than electrostatic copiers. In balance, though, the lack of stable copies and the slow production time are usually more than enough to tip the balance in favor of the electrostatic approach.

Prospective purchasers of copying machines should look for the following:

1. Quality of **image** on the copy, from the standpoint of sharpness, clarity, and **contrast**.
2. Longevity of the copy's stability (i.e., legibility as affected by light, temperature, humidity, passage of time, etc.).
3. Availability of such niceties as a collator, automatic shutting off of the mechanical system, different **enlargement ratio**s, capability to **duplex (copying machines)**, speed (not just in providing the first copy, but in making multiple copies).
4. Power consumption, or energy efficiency.
5. Portability, or movability when required.
6. Cost of supplies (expendables, such as **toner** and paper).
7. Ease of routine servicing and **maintenance**.
8. Rate of **downtime**.
9. Ability to do full color copying, or accurate tonal rendering of colors in shades of black and gray.

As with all other types of equipment, remember Rosenberg's law:

> You may get what you pay for, but the more you get, the more and longer you will continue to pay for all the things you didn't really need. Socratically stated: Don't let the salesman sell you a Peterbilt to haul a pound of nails.

Microform Equipment

With the burgeoning of printed material, which had its last great impetus in the mid-1950s, libraries, schools, government agencies and other institutions which have to store such materials are turning with greater frequency to **microform**s. Devised by John Benjamin Dancer, an English physicist and astronomer, **photomicrography** began on daguerreotype plates in 1939. The **enlargement ratio** was one hundred sixty to one, since Dancer's first **microphotograph** reduced a document twenty inches in length to an **image** only one-eighth of an inch.

The French photographer Rene Dagron applied this technique to the transmittal of documents (using carrier pigeons) during the Prussian siege of Paris, 1870-1871. The technology concerning, and use of, microforms continued apace and, ultimately, resulted in the three basic formats in use today: **microfiche, microfilm** and **microcard**s. All three are, in their initial stages, produced in essentially the same fashion: by photographing a document, page, etc., in such a way that it fills either a portion or the entire **frame** of a piece of **film**. Microfiche and microfilm are similar in that the miniaturized images are on a transparent **base**, i.e., light can be transmitted through the base and the "white" portion of the image. The white portion is the background in a **positive** image and the **print** in a **negative** one. Where microfiche and microfilm differ is in their physical formats: microfiche are sheets, usually 4x6-inches, while microfilm is ordinarily in the familiar **roll film** format, and is either 35mm or 16mm in frame size.

The microcard is an obsolescent format in which the image is on opaque photographic paper. While microfiche and microfilm may be read by the naked eye via a device (reader) which illuminates the image by transmitted light and then enlarges and projects it, microcards are read by devices which must use reflected light. Because of various technical difficulties, this reflected light system does not lend itself as easily to providing a **hard copy** as do the transmitted light systems.

Probably the greatest impetus to the use of microfilm was World War II. During the course of the war, microfilm became useful for a number of purposes: the blueprints of naval vessels and other equipment were more easily transported and stored; V-Mail—letters to and from servicemen overseas—consisted of microfilm reductions of these epistles which allowed, again, for easy transport; and the transmittal of classified information as microdots (a typewritten page reduced photographically to the size of a typewritten period) made covert **information** transfer much safer and simpler.

As the technologies applicable to microforms improved, so did the applications. Banks used them to store such records as checks and other transactional materials. Engineering drawings, blueprints, etc., were stored in microform formats by industry. Hospitals used the **aperture card** (a **Hollerith card** with a 35mm film frame placed in a hole in the card) to store X-rays of patients. (It should be noted that the aperture card is one of the least successful microform formats since, although theoretically the cards can be run through a **card sorter** to locate required films, after just a few such "sorts" the film frames have a tendency to separate from their host cards and seek a new environment: often in the bowels of the sorting device.)

Computer-output-microforms **(COM)**, that is, either microfiche or roll film generated automatically by a **computer**, have become prominent in the last several years. These are of particular use in providing ephemeral statistical data, parts lists, and other such materials which are required quickly but which have relatively short spans of utility.

As for the content appropriate to the two basic formats, i.e., microfilm and microfiche, each has its own peculiar qualities. The microfilm is most useful for storing such serial materials as newspapers, journals, and magazines wherein the material does not need to be **updated** frequently, if at all. Microfiche, on the other hand, being a unitized format, is handy for short reports, catalogs, or material which requires frequent updating; the outdated microfiche can be discarded.

The most recent form of microfiche (i.e., differing from the common **diazo film** or **silver film** types) is the "ultrafiche" which uses a **vesicular film** type material (often known as **photo-chromic-micro-image (PCMI)** and which allows for the storage of approximately three thousand frames on a 4x6-inch sheet (as opposed to the usual ninety-eight frames for "ordinary" microfiche). Because of the prohibitive cost of making an ultrafiche master and "fixing" it (to insure stability of the film), vesicular fiche are not found in the same profusion as ordinary microfiche.

24 / *Criteria for Equipment Selection*

Today, microfiche has become a common information **storage** medium and is issued by such government agencies as the Government Printing Office, the National Technical Information Service, the Educational Resources Information Centers, the National Aeronautics and Space Administration, plus private organizations; and "fiche" are now produced in the tens of millions annually in the United States alone, let alone throughout the world.

All microforms, be they silver film, diazo, or photo-chromic-micro-image, should be stored in an environment whose temperature and humidity are constant and which shields them from light. The storage temperature range should be constant and from sixty-five degrees to eighty degrees Fahrenheit; the relative humidity should constantly approximate a point in the range from 40 percent to 50 percent.

COMPUTERS

Hardware

A **computer** is an electronic device (usually with some mechanical parts) which can, by means of rapid computations, calculations, comparisons, etc., provide solutions to problems—all without human intervention once it is properly **program**med and is fed the appropriate **data**. Contrary to some popular notions, computers neither "think" nor "feel" as humans can.

Computers are of two basic types: **analog** and **digital**. The analog computer represents, measures, and manipulates discrete data (such as individual numbers) by means of such physical variables as voltage changes or step-down in mechanical gears. Slide rules and speedometers demonstrate analog functions. Notice that representations within the limits of such devices are within an infinity of possibilities. Digital computers, on the other hand, like the **calculator**, adding machine, and abacus, deal with discrete data as such on a "quantized" basis. One of the basic differences between analog and digital computers is that the analog can manipulate and represent data directly, while the digital must have the data transformed into some **code**.

Almost all computers are divided into smaller subsystems: **memory, arithmetic unit, control unit, input/output**, etc.

There are three size classifications of computers based on memory and computational capability: **microcomputers, minicomputers,** and **main frame** computers.

A microcomputer is at the small end of the computer size spectrum, having a main memory capacity of less than 500,000 **characters** or **byte**s. A subcategory of the microcomputer, of growing interest to individuals, small businesses, and educational institutions, is the personal computer, with a main memory capacity of 1,000 to 100,000 bytes.

Microcomputers compare and contrast with programmable **calculator**s and larger computers in the following ways.

Calculators are programmable only in sequences of keystrokes, not in **programming language** commands; their **main storage** is limited to a few hundred keystrokes internally, and to postage-stamp-sized magnetic cards externally; they generally manipulate codes for numerals only, not codes for alphabetic characters (although they may be capable of printing alphabetic characters).

Microcomputers and personal computers have much more main memory capacity, or storage, as defined above, than calculators; are programmable in **BASIC** or other **high-level languages** and **assembly language**; utilize **magnetic tape** or **floppy disk** storage to maintain large volumes of data outside of the computer itself. Computers in this class can at present execute only one program, and support only one user at a time.

Minicomputers are equipped with 500,000 to almost 1,000,000 bytes of main storage, can support multiple users, and can execute more than one program simultaneously.

Main frame or large computers have more than one million bytes of main storage, and may be able to support dozens of programs and hundreds of users simultaneously.

Progression from personal, through intermediate sizes, to main frame computers generally means an increase in computational speed; that is, a larger computer will usually execute a program of given size and complexity in a shorter length of time than will a smaller computer. The maximum amount of main storage which a computer can support is limited by **word length**, just as a street can have only one hundred household addresses if house numbers are limited to two digits. In order to **read** from or **write** to a memory location, the computer must be given the specific address of that memory location. The maximum number of addresses available is dependent upon the number of **bit**s in the computer's word. Word lengths vary from four bits for some very small computers currently available to thirty-two or more bits on large main frames. Most minicomputers use sixteen-bit words, while the norm among microcomputers is an eight-bit word.

Software

Current personal and microcomputers are available with computer programs and instruction manuals which permit anyone with high school reading ability to execute prewritten programs or to learn programming. When microcomputers first became available, around 1975, manufacturers concentrated on providing **hardware** and little else. Experienced programmers could use these new, sophisticated "toys" to create programs and perform whatever tasks the talented could envision. More recently, emphasis on the part of manufacturers and vendors has shifted to provision of an immediately useful resource. The computer can be purchased with programs to perform dozens or hundreds of business or household tasks such as checkbook balancing or more sophisticated accounting tasks; tutoring in mathematics, computer programming, or foreign languages; management of building heating and cooling; and just about any other subject imaginable.

In the early phases of microcomputer development, **software** for a given computer was available only from that computer's manufacturer. Now, there is a separate industry composed of companies which produce application programs which are modified to operate on several makes of computer.

A small computer system should be selected primarily on the basis of the task(s) to be performed. A logical first step in the selection process is to determine the availability of software to perform the application in which one is interested. Personal income tax preparation, small business accounting, and programming tutorial for small computers are examples of popular applications.

By examining catalogs, browsing the increasingly common computer section in book stores, or visiting computer vendors, you can learn about software on the market for specific tasks, and about the brand(s) of computer which are capable of executing the desired programs. Determine which brand(s) of computer can run the programs that you have selected, and visit vendors of those computers. A vendor who will allow you to come in and operate a machine which is executing the software in which you are interested, and who will allow you to contact satisfied customers, is one who will likely provide you with satisfactory support.

At one end of the spectrum is the vendor who will provide a **turnkey** system, wherein the seller assumes responsibility for providing everything needed to perform your specified task or application. Others will, competently and with valuable advice, sell only that equipment and those programs which you request. At the other extreme is the seller, often a discount house, capable of providing no advice, but offering equipment to the knowledgeable buyer at a very attractive price.

Determining desired applications and selecting candidate vendors will permit you to estimate cost for a system which will meet your needs. At this writing, a system capable of performing word processing and accounting for a small business can be obtained for as little as two thousand dollars. However, prices for small computers vary so greatly with optional equipment and peripherals, and have been dropping so rapidly (at least 15 percent a year), that any specific price included herein for a system of given capabilities would be obsolete by the time this goes to press.

In selecting a system and vendor, you should not be overwhelmed by promises of infinite capability or by low price alone. The reputations of both the manufacturer and the vendor should be checked, and availability of repair service as well as advice and assistance for customers should be verified. You must be especially wary of claims for the newly available "brand-name compatible" or "look-alike" computers. These imposters are manufactured to physically resemble well known brands of computer, and to use the same programs, while selling for much lower prices. The imposters are frequently inferior in materials, workmanship, and performance; and may be impossible to have repaired if they malfunction.

In determining maintenance availability, determine the extent to which repairs or parts replacement may be made by the user, with simple hand tools, or with no tools at all, and whether or not the vendor has the capability to perform manufacturer-approved repairs on his premises. Service which must be performed at a repair center, especially one which requires shipment of the computer or peripherals, can result in significant **downtime**.

Optional Equipment and Applications

The applications to which the system is to be put will determine which peripheral equipment is required. Many peripherals beyond those required can be nice to have, either for ease of use of the system or for unessential but convenient lesser applications not supported by the required equipment. For instance, word processing applications obviously require a printer (one may create documents with a keyboard, and display them on a **cathode-ray tube**, which is the most common output device for small computers, but one generally types a document for the purpose of eventually creating **hard copy**). **Dot-matrix printer**s are fast and relatively inexpensive, and produce print quality adequate for drafts or informal communication, while formal correspondence would require a slower, more expensive **daisy-wheel** or other full-font printer. By full-font, we mean a printer which, like a standard typewriter, has a one-piece type element for each character the printer can produce.

Special software is available to create and display **graphic** designs on the computer screen. The graphics may be monochromatic or, if the system is equipped with a color monitor, in color. Printers, usually of the dot-matrix type, and **plotters**, are available which can reproduce the screen graphic in monochrome or in color.

Among the most popular and most common applications for microcomputers are education, word processing, the "electronic spread sheet," and graphics.

Computer programs which teach the user are available in a wide range of subject areas, and new subjects become available almost daily. Examples of popular programs are pre-school learning routines that teach pattern recognition, letters of the alphabet, and numbers; math programs that give instruction in single digit addition through calculus; touch-typing programs; and programs that teach computer programming. Software is available for at least one popular computer that teaches reading by use of a digital voice synthesizer which, under computer control, "speaks" to the student.

Word processing is an application which uses to the fullest the capabilities of a small computer, and from which the user can derive great value. With it you can also, through hasty or careless selection of hardware and software, become rather frustrated.

As mentioned above, practical use of word processing software requires selection of an appropriate printer. Word processing also requires relatively large capacity main storage, and disk **auxiliary storage**.

Line display is also an issue. Many microcomputers have, as standard features, hardware and control programs which display thirty-two or forty characters or spaces on each line of the display device. (These microcomputers are designed in such a way that standard television receivers may be used as the cathode-ray tube display.) Since 8½-inch wide paper, used with one inch left and right margins, and twelve-pitch type (12 characters per inch), allows seventy-eight characters per line, "serious" word processing use requires at least that many characters per display line. Some small computer systems present the first thirty-two or forty characters of a text line on one display line, with the remaining characters "wrapped around" onto the next display line. While this display technique is intelligible, it is cumbersome and suited only for the most casual use. A high resolution monitor and eighty-column (in computer parlance, an eighty-character line is an eighty-column line) display are, therefore, necessary for professional quality word processing. Eighty-column displays are the de facto standard for word processing cathode-ray tube displays.

Finally, a few cheap computers are still to be found which display only upper case alphabetical characters. This, too, is a limitation which precludes most word processing use, and certainly precludes production of camera-ready copy.

Word processing programs available for microcomputers possess a wide range of text handling capabilities. One of the most basic is the insert or delete function, to allow single characters or words, or even whole paragraphs or pages, to be added to or deleted from the text, with unmodified text "opened" or "closed" to provide proper spacing. Automatic "carriage return," whereby the typist need not manually execute a return at the end of a line, is another standard feature. When the right margin is reached, the **cursor** automatically advances to the next line, with any partially typed word being simultaneously transferred to the left margin on that line. Some of the more advanced functions are right margin **justification**; the ability to transfer columnar material left or right across a page; and an online dictionary which highlights words which do not match a dictionary entry, and which may provide automatic hyphenation. Of course, the more capable word processing programs are the most expensive, and require the largest main and, possibly, auxiliary storage.

Far less obvious to the first-time purchaser of word processing software is the fact that some programs are much easier to use, or more **user friendly** than others. Each of a program's functions must be invoked by the entry of a command. Obviously, the greater the number of available functions, the greater the number of commands which must be used. The best programs for new and experienced users alike, in the author's opinion, offer a "help" function, which will list the available commands on the cathode-ray tube without disrupting the work in progress. In selecting a particular program, one should ask for a demonstration by a skilled user, in order to quickly determine the availability of desired functions, and then gain some "hands-on" experience with that program.

An application which rivals word processing in utility is the "electronic spread sheet." The spread sheet program gives the user a matrix of "cells," each of which may contain text or numbers. This matrix is typically thirty-five or more columns wide (column here referring to vertical stacks of multi-character cells), and hundreds of lines long. Six or eight columns and sixteen to twenty-four lines of the matrix are visible on the cathode-ray tube at one time; the program permits horizontal and vertical "scrolling" to bring the entire matrix into view one piece, or "window," at a time. The unique feature of the electronic spread sheet is that each cell can be programmed to receive results of calculations performed on the contents of other cells or groups of cells. For example, text can be entered into appropriate cells to create a facsimile of Internal Revenue Service form 1040. Cells in proper locations can be programmed to receive sums, percentages, subtotals, etc., from other sets of cells so that, when the last item of raw data has been entered, all calculations have been performed, and the complete return is ready to print.

The final individual application to be addressed herein is that of graphics. Software is available which permits the automated conversion of tabular numeric data into bar graphs, line graphs, pie charts, and other graphic representations, in several colors if the computer system is equipped with a color cathode-ray tube display. As mentioned earlier, printers and plotters are available which will reproduce displayed graphics as hard copy.

Combined applications software packages (i.e., those which are "interactive" or which "inter-relate") are becoming available for many brands of microcomputers. These packages provide word processing, electronic spread sheet, graphics, and sometimes other features such as filing and information retrieval, in a single set of programs. These programs function in such a way as to share data. The word processor can be used to enter and edit columns of numbers, which are processed within the spread sheet to calculate row and column totals and other derived data, which in turn are processed by the graphics program section to produce diagrammatic analyses—all with the entry of each number but a single time. Correction of a number within any application automatically corrects that number (and any derived data or graphic display dependent upon it) throughout the multiple applications package. This type of capability should be construed as a definite selection factor at the time of purchase, since it provides both speed and, assuming the appropriate data are entered, a greater assurance of accuracy.

A combined applications software package which includes filing and information retrieval capabilities can be of value in management of school or other small libraries, assisting with such routine chores as circulation control, compilation and printing of reference service statistics, and development of selective dissemination of information (SDI) profiles based on the library's collection. When equipped with communications software and a **modem**, the small computer can serve as a computer **terminal**, to access remote computers.

Dictionary of Library and Educational Technology

A/C. *See* **alternating current**.

ADP. *See* **data processing**.

AFC. *See* **automatic frequency control**.

AGC. *See* **automatic gain control**.

ALC. *See* **automatic level control**.

AM. *See* **amplitude modulation**.

APL.
 Initialism for "a **programming language**." APL first came into use in 1967 and was devised primarily for the IBM 360 **computer**. It is a **high-level language** whose applications are essentially suitable for mathematics.

ASA. *See* **American National Standards Institute**.

ASCII.
 Acronym for American Standard Code for Information Interchange. Like **EBCDIC**, ASCII determines the number of **bits** for each **character**. ASCII (also known as the United States of America Standard Code for Information Interchange, or USASCII), although it was originally a seven bit **code**, is now more frequently encountered as a seven bit code embedded within an eight bit code, and referred to as ASCII-8.

ABERRATION.
 The **distortion** of either the shape or color of an **image** caused by some failure in design of a lens or an optical system. When the distortion affects the color of the image, it is technically referred to as chromatic aberration.

ACCESS.
 The ability to locate **data** in a **storage** device and transfer it either to another storage or **input/output** device.

ACCESS TIME.

The time it takes from the instant of entering the command requesting that specific **data** be transmitted from **storage** until the instant the data arrives at the destination storage or **input/output** device.

ACETATE TAPE.

A tape **base** designed for **magnetic recording** which is made of a strong, transparent plastic. The main advantage of acetate tape is that it does not stretch appreciably. However, it has a tendency to age badly: it becomes brittle and, after long use or storage, breaks more easily than **polyester tape**. Acetate tape should be stored in a relatively dry and stable (sixty to eighty-five degrees Fahrenheit) environment. It is not wise to store acetate tape that has been too tightly wound on its **reel**. Store reels of acetate tape vertically, preferably in individual containers.

The use of acetate tape has been essentially discontinued since about 1960. For purposes of long term preservation, material stored on acetate tape should be transferred to polyester tape.

ACOUSTIC COUPLER. *See* **modem**.

ACOUSTIC FEEDBACK.

A phenomenon manifested as a loud howl, caused when sound re-enters the system that generated that sound. It is most commonly encountered either when sound from a **loudspeaker** re-enters the **microphone** in the sound system or when the sound from a loudspeaker causes a **phono cartridge** in the system to vibrate. In the former instance, the volume may be turned down and/or the microphone placed farther from the loudspeaker. In the second case, again, lowering the volume may help. Also, isolating the **turntable** from the loudspeaker and, if necessary, placing the turntable on a wall-mounted shelf may alleviate the problem.

ACOUSTIC SUSPENSION.

A type of **loudspeaker** system design, also known as air suspension. The speakers are mounted in an hermetically sealed **enclosure**. Acoustic suspension usually offers better **bass response** than similarly sized systems of other design. There has been a tendency for acoustic suspension systems to be of **low-efficiency**, but recently acoustic suspension systems of comparatively **high-efficiency** (in the ten to twenty watts range) have been produced. Most acoustic suspension systems have an **impedance** of eight or four ohms. A good **frequency response** for an acoustic suspension system would be from about thirty **hertz** to twenty thousand hertz plus or minus three **decibel**s and a desirable **resonance frequency** would be thirty-five hertz or lower.

ACTIVE MEMORY.

In a **calculator**, a **memory** which, even when the device is switched off, is fed a minute electric current which is sufficient to allow **data** entered while the device was on to be kept in **storage** and which enables the user to **access** and manipulate that data when the unit is again switched on.

ACUTANCE.

The measure of a lens, or **film**, or both, which indicates the sharpness of the line between high- and low-exposure areas. Commonly accepted as a measure of "acuity" or sharpness.

ADAPTIVE PROGRAM. *See* **programmed instruction**.

AIR SUSPENSION. *See* **acoustic suspension**.

ALGORITHM.
A series of steps, usually based upon some known rule, followed in order to solve a problem. In work with **computer**s, an algorithm is ordinarily expressed in such a way, in a suitable language, as to be comprehensible to the devices employed.

ALIGNMENT, HEAD. *See* **head**.

ALPHAMERIC. *See* **alphanumeric**.

ALPHANUMERIC.
A **character set**, or expression, which includes at least alphabetical and numerical **character**s, and may well include others (punctuation marks, arithmetic operators, etc.).

ALTERNATING CURRENT (A/C).
A flow of electricity that cyclically reverses its direction. In the United States, the cycle takes place sixty times a second and the current is said to have a sixty **hertz frequency**. Alternating current is the type usually supplied by electric companies to their customers. Dry cells (and other batteries) ordinarily produce **direct current**.

AMBIANCE.
A term used in one of two senses. In the first sense it describes the conditions of the environment in which a live performance may have taken place — e.g., concert hall ambiance. The second sense refers to the condition of the playback environment — e.g., listening room ambiance. A former goal had been to reproduce the concert hall ambiance within the ambiance of the listening room by means of improved equipment, etc. This appears to be a rather hopeless aim because of the impossibility of making a listening room into a concert hall.

AMBIENT LIGHTING. *See* **natural light**.

AMERICAN NATIONAL STANDARDS INSTITUTE (ANSI).
This organization began in 1918 and from that date to 1928 was called the American Engineering Standards Committee. From 1928 to 1966 it was known as the American Standards Association (ASA); from 1966 to 1969 its title was the United States of America Standards Institute. In 1969, the present title was adopted. The function of ANSI is to set design and production (and even some functional) standards for a huge variety of products manufactured in the United States. For many years, the sensitivity to light of **film** manufactured in the United States has been specified as an ASA rating. Despite the title changes of the organization, this specification of **film speed (sensitivity)** has continued (by both manufacturers and users) to be given as an ASA number.

The closest equivalent to ASA numbers for film are the film speed rating numbers assigned by the **International Standards Organization (ISO)**, whose numbers accompany most film manufactured in Europe and Japan (and are also provided by most American manufacturers on their film products).

The address and telephone number of ANSI are: 1430 Broadway, New York, New York 10018; (212) 354-3300.

AMERICAN STANDARDS ASSOCIATION. *See* **American National Standards Institute**.

AMMONIA PROCESS. *See* **diazo film.**

AMPLIFIER.

A device that increases the **amplitude** of a **signal** fed to it. Since the mid-1960s, amplifiers have primarily been **solid-state devices** and have not used vacuum tubes.

An **audio** amplifier enlarges the small **input** signal from the **tuner, phono cartridge,** or tape **deck** and then supplies that signal to the **loudspeaker.** Most modern audio amplifiers tend to be integrated amplifiers, that is, they incorporate a **preamplifier.** Specifications appropriate to audio amplifiers involve **channel separation, frequency response, hum, intermodulation distortion, power bandwidth, power output, rated power output,** and **signal-to-noise ratio.**

Among the inputs that an integrated audio amplifier should have are **auxiliary input; phono input,** usually for **magnetic cartridge; tape input,** usually for tape deck; and **tuner input.**

Basic audio amplifiers, that is those designed without a preamplifier on the same chassis (and sometimes called power amplifiers) usually have no controls. An integrated amplifier usually contains the following controls: **balance; input selector;** loudness compensation (*see* **Fletcher-Munson Curves); mode selector; monitor; tone controls;** and **volume control.**

Two other controls that are occasionally found on such equipment and that are desirable are the **filter switch** and the **phase reverse control.**

A **video** amplifier is one that takes the small electrical signal from the **output** of a television camera or the **video head** of a **videotape recorder** and amplifies it for viewing on a video monitor. The video amplifier works in essentially the same fashion as an audio amplifier.

Important specifications appropriate to video amplifiers are **bandwidth** and signal-to-noise ratio.

AMPLITUDE.

The designation of the size (i.e., height) of a **signal,** as distinct from its **frequency.** In **audio** equipment, the term is often used interchangeably with volume and refers to how loud a given signal may be. Audio amplitude is usually measured in **decibels. Video** amplitude is a measure of the brightness of a television picture.

AMPLITUDE MODULATION (AM).

This usually refers to a radio broadcast system in which **modulation** is accomplished by varying the **amplitude** of a fixed, high **frequency signal** called the **carrier.** The AM radio **band,** in the United States, covers the assigned broadcast frequencies of 535,000 **hertz** to 1,605,000 hertz. Because some forms of natural and man-made electrical disturbances are also a form of amplitude modulation, these are sometimes received as static on AM **receivers.** Also, the AM band, when compared to the **frequency modulation** (FM) band, is not as good for **high fidelity** reproduction, being usually limited to a **frequency range** of about 100 hertz to 8,000 hertz. One advantage of AM broadcasting is that it covers a greater distance than FM, because AM waves follow the curvature of the earth and FM waves do not.

ANALOG.

The means of representing **data, information,** or physical quantities by other, variable, physical quantities. For example, the grooves in a **phonograph disc** contain waveforms that are analogous to the waveforms (in space) of the sound therein represented. In a large sense, analog may be thought of as metaphoric while **digital** (with

ANALOG (cont'd)
which analog is often contrasted) might be imagined as the quantification of real entities (including sound and information).

ANALOG COMPUTER. *See* **computer**.

ANECHOIC CHAMBER.
Ideally, a room in which there are no surfaces (walls, ceiling, etc.) that could reflect sound waves. **Audio** engineers use anechoic chambers that approximate this ideal to test such devices as **loudspeakers** and **microphones**.

ANIMATION, FILM.
A process that simulates motion (used, for example, in animated cartoons) by photographing a series of pictures or drawings, each showing a stage of movement slightly changed from the drawing (**frame**) before, so that the **film** achieves the illusion of motion when these frames are shown sequentially.

ANTENNA.
Metal rods, wires, etc., used to pick up radio waves that have been broadcast into the atmosphere. There are various types of antennas, most of which are directional. This means that they are more receptive to **signal**s emanating from a given direction; therefore, directional antennas must be "aimed" in order to achieve best reception. Good antennas are usually devised to receive a specific type of wave, so it makes good sense to match the antenna to the type of wave to be received (**frequency modulation**, television, shortwave, etc.). The antenna is usually connected to the device it serves by means of a **lead-in**, often **coaxial cable**. To prevent damage to the **hardware** to which an outdoor antenna is connected, the antenna should be properly grounded.

ANTI-SKATING.
This term refers to some method of overcoming a mechanical phenomenon found in most **phonograph turntable**s. The phenomenon (skating) occurs when the **stylus** is pulled toward the center of the **disc** by the groove of the disc it is tracking. This pulling, or skating, causes some **distortion** in playback, and better turntables provide a means of preventing this phenomenon, usually in the form of an anti-skating counterbalance, a magnetic counter-force, etc.

APERTURE.
An opening in a camera or projector that allows for the passage of light through the lens. In cameras, the size of the aperture is adjusted by the **iris**. This allows for a variance of the amount of light striking the **film** and also affects the **depth of field**. In projectors, the aperture allows light to flow through the **image** to be projected and also properly positions that image. The size of the aperture relative to the **focal length** is given as an f number, which indicates **lens speed**.

APERTURE CARD.
A card (usually a **Hollerith Card**) with a rectangular hole cut in it; into which may be fixed a piece of **microfilm** (ordinarily on a frame of 35mm **film**). At one time aperture cards were thought to be an excellent means of **information storage** and **information retrieval** for such things as large engineering drawings and X-ray films. The principal problem encountered in their use was that, as the cards were run through a **card sorter**, the microfilm tended to separate from the cards. Aperture cards are still to be found in use in

APERTURE CARD (cont'd)
some medical facilities to store X-ray film, but are not usually manipulated other than manually.

ARC LIGHT.
Type of **artificial light** produced when electric current is passed through a carbon rod the point of which is separated from the point of another carbon rod by an air space. The light produced is a result of a spark which bridges the gap between the two rod tips. Arc light tends to be of high intensity (i.e., quite brilliant) and is used most often in the production of motion pictures.

ARCHIVAL STANDARDS.
A group of standards promulgated by a number of organizations (**American National Standards Institute**, Society of American Archivists, etc.) which set minimum quality levels for the specific longevity of a variety of **information** records (**microform**s, **magnetic tape**, paper, etc.).

ARITHMETIC UNIT.
The portion of a **computer** which carries out operations on numbers (e.g., addition, division, subtraction, multiplication, comparison of "greater or less than") and is a part of the **central processing unit**.

ARTIFICIAL LIGHT.
Light other than **natural light** (i.e., light produced by the sun or stars, or sunlight reflected by the moon, etc.). There are two basic sources of artificial light: combustion (e.g., the burning of matter such as in candles, flares, or torches) and electrical (of which there are essentially four types: neon; **arc light, fluorescent light,** and **incandescent light**).

Natural light is usually desirable for taking aesthetic photographs (portraiture, landscapes, and the like). Artificial light, being more controllable, is the type ordinarily employed for the **exposure** of **microfilm** or other photographic records.

ASSEMBLY LANGUAGE.
Computer language which is supplanting **machine language**. Assembly language depends upon the computer's ability to translate that language's terminology (which is usually based upon sets of mnemonic devices, etc.) into machine language. Generally, **program**s written in assembly language are faster and easier to write and more easily interpreted by humans.

ASYNCHRONOUS TRANSMISSION.
Transmission of **data characters** in which each character is preceded by a "start element" and followed by a "stop element." In this way, each character is separated from adjacent characters.

ATTENUATION.
The diminishing in strength (**amplitude**) of a **signal**—**video**, electrical, **audio**, etc.—usually measured in **decibels**.

AUDIBLE FREQUENCY RANGE.
The range from the lowest **frequency** to the highest, capable of being heard by the human ear. This range is normally from about forty **hertz** to about eighteen thousand hertz. As a person ages, the upper portions of this range may decrease, usually down to about eight thousand to twelve thousand hertz.

AUDIO.
As it pertains to television and motion picture **film**s, this designates the sound portion of the **program** as distinguished from the **video** or picture portion.

Audio may also refer to a component in a sound system.

AUDIO CARTRIDGE. *See* **cartridge, audio.**

AUDIO CASSETTE. *See* **cassette, audio.**

AUDIO-VISUAL AIDS. *See* **instructional technology.**

AUDIO-VISUAL MATERIALS. *See* **instructional technology.**

AUTOMATIC CAMERA.
Camera in which any of a number of functions are done automatically. Among the most common automatic functions are focus, **shutter speed** selection, advancing the **film** to the next **frame**, and setting the size of the **iris** aperture. Automatic focus is sometimes called automatic lens.

AUTOMATIC DIALER.
Device which has a small **memory** in whose **storage** several telephone numbers may be entered. In most such units, the user presses but one button and the mechanism then emits the appropriate **rotary dial** impulses or **tone generated dial** tones. Automatic dialers are, obviously, most useful for storing frequently called numbers.

AUTOMATIC DUPLEX. *See* **duplex.**

AUTOMATIC FEED.
In **facsimile transmission**, a means for providing continuous and automatic insertion into the system of those items to be transmitted. Usually the materials are on separate sheets, each fed into the transmitter as soon as the previous sheet has been scanned and transmitted. Automatic feed has the obvious advantages (over manual feed) of running both faster and unattended.

Automatic feed, also called batch feed, is often employed in **copying machines** as an adjunct to **document feed**, and in **word processor**s for uninterrupted printing of large documents.

AUTOMATIC FOCUS.
Camera or projector feature which, without human aid, correctly sets the focus of the lens. This is usually accomplished in one of two ways: either by measuring the elapsed time for an inaudible, **ultrasonic signal**, generated by the camera or projector, to reach the photographic subject or **screen** and return to its source; or by "reading" the distortion of the subject or projected **image**. In either case, the adjustments necessary to effect correct focus are virtually instantaneous.

AUTOMATIC FREQUENCY CONTROL (AFC).
A system used in many **frequency modulation** (FM) radios, **tuner**s, etc. The automatic frequency control is a circuit built into such devices to prevent the station to which it is tuned from **drift**ing. Better tuners, etc., do not need AFC, but those devices which do use AFC should offer a means of "defeating" the AFC since, on occasion, the AFC will not lock onto a relatively weak station adjacent to a strong one, and since it may cause a loss of some upper **frequency** portions of the material being broadcast.

AUTOMATIC GAIN CONTROL (AGC).

This is similar to **automatic level control**, but it generally refers to use in **videotape** recording and television. In such devices it is a circuit that maintains the **amplitude** of the picture **signal** at a given level—usually for recording purposes. **Videotape recorder** cameras equipped with AGC ordinarily need no operator compensation for changes in the light conditions, since these changes are made electronically and automatically. It is helpful, in devices using AGC, if the automatic feature can be "defeated" or switched off so that human judgment may be allowed to determine appropriate levels, particularly under extreme conditions. AGC, however, does simplify the operation of equipment in which it is found because it enables relatively inexperienced personnel to operate otherwise complicated devices.

AUTOMATIC LENS. See **automatic camera**.

AUTOMATIC LEVEL CONTROL (ALC).

This is analogous to **automatic gain control** and usually refers to **audio** equipment. Most commonly, an ALC circuit is found in **cassette** recorders. Here it is used to maintain an optimal audio recording **level**, so that the operator does not need to change the recording **volume control** as the **amplitude** of the material being recorded changes. Better equipment that incorporates ALC will also provide a means for "defeating" it, or switching it off, so that the user can adjust the recording level, particularly in exceptional situations. ALC does simplify the process of making a recording, but it should be used primarily when recording speech, since it tends not to respond quickly enough to the varied amplitude changes of music. Also it will tend to average or "iron-out" the musical amplitude differences to a median level.

AUTOMATIC RECORD CHANGER. See **turntable**.

AUTOMATIC REPLAY.

In sound or visual systems (usually those employing **magnetic tape** or **phonograph discs**), a system for replaying a preselected portion of the **program**. In some systems this is accomplished by means of a **signal** recorded on the **medium** or, in others, by presetting a mechanical device.

AUTOMATIC REVERSE.

A method of changing a **magnetic tape**'s direction of travel during playback or recording. In **open-reel** magnetic tape devices, the automatic reverse usually operates in the playback **mode**. There are two basic methods of automatic reverse in open-reel machines. The most common uses a foil sensing tape applied to that point of the tape at which reverse play is desired. Some of these devices are able to effect **foil tape sensing** at both ends of the tape; when switched to an automatic mode, they will continue to play a prepared tape, first in one direction and then the other, until the operator of the device intervenes. The other means of accomplishing automatic reverse is by recording a specific **signal** (usually low-**frequency**, in the vicinity of twenty **hertz**) at the point on the tape where reverse play is desired. In this system, too, some machines are equipped to play a tape continuously, first one side and then the other.

In audio **cassette** devices, reverse play is also possible in two ways. One reverses the direction of the tape when a sensing device senses cessation of the tape motion. The other method uses a mechanical device that physically turns the cassette over when the sensing device indicates that the tape has run through completely.

AUTOMATIC REVERSE (cont'd)
Systems that play the open-reel tape or the cassette in both directions (without turning the cassette over physically) usually need two **playback head**s, one for each direction of play. The same is true for devices that are able to record in both directions—i.e., two record heads are ordinarily required.

Automatic reverse is an excellent feature in this type of equipment because it allows a complete program to run automatically. Such a feature has numerous institutional applications. The most obvious of these is that it can provide a continuous **audio** program to run with a **slide projector** or **filmstrip** projector that can be operated by remote control and that is connected, by means of a **synchronization** device, to the recorder. The tape, of course, must have been preprogrammed to cue the projector when the projected **image** is to be changed.

AUTOMATIC REWIND.
A method employed in various types of **tape recorder**s to **rewind** the tape from the **take-up reel** to the **supply reel**, once the tape has been played through in the normal, left-to-right, fashion. Lately, some **motion picture projectors** have also incorporated automatic rewind capabilities. Tape recorders (usually **open-reel** types) that offer automatic rewind ordinarily accomplish it by means of a **foil tape** sensing system applied near the end of the tape.

Automatic rewind for projectors is frequently effected in the following manner: the end of the **film** to be shown is locked to the hub of the supply reel; when the projector senses the drag on the supply reel's spindle, the automatic rewind **mode** is activated.

AUTOMATIC SEARCH.
Means of finding a specific portion of a **magnetic recording** in which the **magnetic tape** is transported in either the **fast forward** or **rewind mode** at which point the device then stops. Some devices will then, automatically, place themselves in the play mode.

AUTOMATIC SHUTOFF.
A feature that is found in many devices, but principally in three basic types of equipment: **tape recorder**s; **motion picture projectors**; **slide projectors**; and **phonograph turntable**s. When the tape, **disc**, etc., has been played or shown to its conclusion, the device shuts off its motor and possibly its electronic devices (such as **amplifier**s).

AUTOMATIC TURNTABLE. *See* **turntable**.

AUTOTHREADING. *See* **motion picture projectors**.

AUX. *See* **auxiliary input**.

AUXILIARY INPUT.
This refers to the **jack** on an **amplifier, preamplifier, receiver, tape recorder**, etc., into which the user may connect another device. For example, assuming that the auxiliary **input**s and the **output**s of the secondary device are matched in **impedance** and power, one may plug an audio cassette **deck** into the input jack of an **audio** amplifier in order to feed the **cassette** deck's **signal** through the system powered by the amplifier. Most United States manufacturers design equipment with an auxiliary (or high-level) input able to accept an impedance of one hundred thousand ohms or more, and audio signals of approximately one-tenth to one volt.

AUXILIARY STORAGE.

Storage of **data** in some **medium** other than **main storage** (e.g., **magnetic tape, punched card, buffer**). Auxiliary storage is used because it is usually cheaper and of greater capacity than main storage, but auxiliary storage is almost always characterized by slower **access time**.

AVAILABLE LIGHT. *See* **natural light**.

BASIC.

Acronym for Beginner's All-purpose Symbolic Instruction Code, BASIC is a **high-level language** designed (about 1967) to be simple and inexpensive and to be used as an **online programming language**. It is especially popular in home and academic environments.

BPI.

Abbreviation once used for **bit**s-per-inch and now, more commonly used to mean **byte**s-per-inch. BPI primarily refers to **packing density**.

BPS.

Abbreviation originally taken from **bit**s-per-second and now used for **byte**s-per-second. BPS is used to describe the transmission rate of **digital information** or **data** described by **baud**.

BACK PROJECTION. *See* **rear screen projection**.

BAFFLE.

In **loudspeaker** terminology this refers to the board on which the speakers are mounted—so called because it separates (or baffles) the sounds being generated from the front and rear of the speaker. To a certain extent, it alleviates the cancellation of some sounds emanating from the front of the speaker, particularly in the **bass** range.

There are various types of baffles. Two of the most important are the infinite baffle and the bass reflex (or ported or vented) baffle. In the former, a very large amount of space is provided behind the loudspeaker either by mounting the baffle on the front of a very large cabinet or by using a wall, door, etc., as the baffle, with an entire room or closet as the enclosed space to the rear of the loudspeaker. The bass reflex system uses a hole (or port) in the baffle. This hole, whose dimensions are critical and related to specifics of the loudspeaker (size, position, etc.) and of the **enclosure** itself (size, **resonance frequency**, etc.), allows sound waves coming from the rear of the loudspeaker to exit through the enclosure without interfering with sound waves emanating from the front of the loudspeaker.

BALANCE.

A term used in reference to **stereophonic** or **quadraphonic audio** systems. Balance has been achieved when all **loudspeaker**s in such a system are fed an appropriately loud **signal** in order to approximate the original **ambiance** of the environment in which the material was recorded. Stereophonic systems require left-to-right balance, and quadraphonic systems need front left-to-right and rear left-to-right balance, plus front-to-rear balance. Many systems use individual **volume control**s for each **channel**. However, once all the loudspeakers have been balanced, a master volume control provides a wider range of overall amplitude.

BALOPTICON. *See* **opaque projector**.

BAND.
A group of frequencies within two limits, usually designating a specific radio broadcasting "section" assigned by the **Federal Communications Commission**. For example, the complete short-wave band is thought of as being from 1,600,000 **hertz** to 30,000,000 hertz.

A **frequency range** is similar, but it is not ordinarily used to describe broadcasting **channel**s.

BANDWIDTH.
This ordinarily means a stated **frequency range** at rated **output** and within a rated **distortion** tolerance. Thus, the bandwidth of an **amplifier** may be twenty **hertz** to twenty thousand hertz, at thirty watts per **channel**, with a .04 percent **harmonic distortion**.

Bandwidth is most often applied to amplifiers and **receivers**.

BAR CODE.
A method of encoding **data** by using vertical bars and spaces. The most common form currently in use is on consumer goods found in supermarkets. The bar code is **input** by passing it over an **optical character recognition** device. There are several standards governing bar code design, the most common being the Universal Product Code (UPC) and CODABAR.

In libraries, bar codes are used in circulation, among other systems, because of the speed with which data can be read into the **computer** and because their use eliminates the need for **keypunch** operations.

Bar codes are sometimes referred to as zebra codes because of their appearance.

BASE.
The backing for **magnetic tape** or **film**. The base (for magnetic tape) may be either **acetate tape** or **polyester tape**. The base, almost always transparent, should provide a flexible, yet tough, **medium** for the magnetizable material bonded to it. The magnetizable material (usually an oxide of some sort) gives magnetic tape its characteristic brown or blackish color and is termed the **coating**.

Magnetic tape comes in various thicknesses, depending on the thickness of the base itself, measured in **mils** (thousandths of an inch). Tapes generally come in thicknesses of ½ mil, 1 mil, and 1½ mil. The thicker the base, the less chance there is of **print through** and **pre-echo**. However, the thicker the base, the smaller is the amount of tape that can fit on a **reel** of a given size, thus providing relatively less playing time.

At one time motion picture film was made of cellulose nitrate. It is now made of the less explosive and less flammable cellulose acetate.

BASIC AUDIO AMPLIFIER. *See* **amplifier**.

BASS.
The sounds in the lowest **audible frequency range**, from about 40 **hertz** (a sound that can only be felt as vibrations in the floor) to about 200 hertz (middle C on the piano is 256 hertz). Bass sounds tend to be omni-directional—that is, they appear to emanate from no specific source—while **treble** sounds tend to be highly directional. The **loudspeaker** component usually associated with the reproduction of bass sounds is the **woofer**.

BASS REFLEX. *See* **baffle**.

BASS RESPONSE.
The extent of the ability of a device to reproduce the **bass frequency range**. A good **loudspeaker** should be able to reproduce sounds down to about fifty **hertz** with little **distortion**. An **amplifier** (or **receiver**) should be able to reproduce bass frequencies down to about twenty hertz or less with virtually no distortion.

BATCH FEED. *See* **automatic feed**.

BATCH PROCESSING.
The processing, as a single group, of accumulated and prepared **data**, of a similar nature, by a **computer**. Although the preparation may be time-consuming, and the time of processing may be delayed to take advantage of equipment when it is least used, the economies of batch processing are often thought to balance the delays. Batch processing is often regarded as less advantageous, however, than **online** processing.

BAUD.
Named for the inventor J. M. E. Baudot, baud, also called **bit** rate, is a unit of **data** communications transmission. In its simplest definition, baud is taken to mean **bytes**-per-second (**BPS**). However, complexity enters when, in multi-state **signal**ing, baud is equal to the number of discrete signal events per second. In common parlance, once again, the baud rate usually approximates bytes-per-second.

BAUD RATE. *See* **baud**.

BEADED SCREEN. *See* **screen**.

BEGINNER'S ALL-PURPOSE SYMBOLIC INSTRUCTION CODE. *See* **BASIC**.

BELT DRIVE.
A system that provides rotational motion from the motor to another part of that system by means of an endless flexible belt (usually rubber or some polymer plastic). It is not the most desirable type of drive system, particularly in **tape recorder**s, **phonograph turntable**s, and **motion picture projectors** because the belts ultimately tend to stretch and thus to rotate either at an improper speed or not at all. The **direct drive** system is preferable, particularly in turntables and tape recorders. Either the belt drive or the direct drive system is definitely preferable to the **idler wheel** system in turntables.

Most motion picture projectors are of the belt drive type, but some of the better ones use a gear drive (or direct drive) system which, although potentially noisier, eliminates the problems encountered with belts.

BENCHMARK.
A specified task, representative of a larger piece of work, by means of which the ability of a person, company, or piece of equipment can be measured in the event that the person, company, or piece of equipment is assigned the larger endeavor. In short, a practical test.

BETAMAX.
The registered trademark of the Sony Corporation for its popular **video cassette**-style **tape recorder**. Betamax cassettes use a tape which is one-half inch wide and which has a playing time that varies from thirty minutes to five hours. Several other companies have been licensed by Sony to produce devices using the Betamax (or Beta) format. In competition with Betamax is the **VHS** format of the Victor Company of Japan (JVC).

Beta cassettes are designated by the length (in feet) of tape they contain. For example, an L750 cassette contains a tape 750-feet long and has a playing time of one hundred minutes at the fastest Beta speed (which speed is no longer incorporated in new Beta machines except for playback, not record) and 4½ hours at the slowest (and newest) speed. As regards quality differences between Betamax and VHS formats, there are essentially none.

BETWEEN-THE-LENS SHUTTER.
A kind of **shutter** most often encountered in the **optical viewing system** type of camera. Sometimes called a **diaphragm** shutter, it consists of a ring of three or more overlapping metal blades located within the lens system itself. These blades are usually closed but can be opened for a relatively precise amount of time. When open, the blades allow light to pass through the lens, thus exposing the **film** within the camera. The opening formed by the blades is usually star-shaped. The between-the-lens shutter ordinarily has a range of speed settings (i.e., timings during which the shutter remains open) of from 1 second to 1/500 of a second. These timings are called **shutter speed**s.

In the **single lens reflex camera**, the usual type of shutter used is the **focal plane shutter**. This kind has the advantage of permitting faster speed settings (to 1/1000 of a second and faster) and more accurate speed timings.

BIAMPLIFICATION.
A means of improving the sound quality of an **audio** system by first dividing a **signal** into **frequency band**s (by means of a set of **filters**), then feeding each band to an **amplifier** specifically designed to handle that band. Each amplifier, then, delivers its amplified signal to an equally specifically designed **loudspeaker**. The term *biamplification* came into use when only two amplifiers were used, each of which "drove" a **tweeter** or **woofer**. In the more complex modern audio systems, where each **channel** may use three or more **drivers**, a more accurate term would be *polyamplification*.

BIAS.
A fairly high (70,000 **hertz** to 120,000 hertz) **signal** recorded in the **magnetic tape** recording process along with the regular **audio** signal. The purpose of recording this high signal is to obtain a more even recording (i.e., one that has less **distortion**).

BIDIRECTIONAL PLAYBACK.
A system frequently found in **tape recorder**s, in both the **open-reel** and the audio **cassette** types. Such systems usually provide two **playback head**s, which allow the tape to be played in both directions without turning the tape over to play the second side. Bidirectional playback is ordinarily associated with **automatic reverse**.

BIDIRECTIONAL RECORD.
A system in **tape recorder**s that allows for recording (usually a **stereophonic** program) on both sides of a tape without turning the tape over to record the second side. It is usually accomplished by using two **record head**s—that is, one for each direction of tape travel. Bidirectional record is frequently found on machines offering **automatic reverse**.

44 / BINARY

BINARY.
A numerical system which uses a base number of two (as compared with the common decimal system whose base number is ten). There are two digits in the binary system: one and zero, but, in binary notation, all other numbers can be represented by these two digits.

Because a **computer** can only distinguish between "off" and "on" (or an electrical impulse or the absence of that impulse) binary notation is the usual form of coding **data** for computer manipulation. (This is not to say that other numerical systems are not used also, e.g., octal and hexidecimal.)

As an historical note, the binary system was devised by the German philosopher-mathematician G. W. von Leibnitz as an attempt to prove the existence of God ("unity" or "one") and all else (the "void" or "zero"). Leibnitz also improved the calculating device of Blaise Pascal, enabling it to multiply and divide directly without imposing the use of an **algorithm**.

BINARY DIGIT. *See* **bit**.

BINAURAL PHONOGRAPH. *See* **Cook System**.

BIT.
Contraction of **binary** digit, bit is the smallest unit of **data** to be manipulated by a **computer**. It is usually taken to mean either zero or one in binary notation.

BIT DENSITY. *See* **packing density**.

BIT RATE. *See* **baud**.

BITS-PER-INCH. *See* **BPI**.

BITS-PER-SECOND. *See* **BPS**.

BLACK BOX.
A device (so called because, in the past, of its housing) which is usually connected to another system, or systems (ordinarily of greater complexity than the black box) and which enables the other system(s) to perform either additional functions or functions different from those originally intended. More often than not, the user need not know how the contents of the black box operate in order for the enclosed device to be used properly.

BLACK LIGHT.
Invisible light, usually **ultraviolet**, which is important in certain types and aspects of photography.

BLACKBOARD. *See* **chalkboard**.

BLADE SHUTTER. *See* **between-the-lens shutter**.

BLOCK.
A set of contiguous **data** records or words which are handled as one unit by a **computer**. Either end of a block is separated from adjacent blocks by a **delimiter**. On **magnetic tape**, the delimiter is usually an inter-block gap.

BLOWBACK.
As a verb or a noun, the act or product of an **enlarger** from a **microform**. A blowback is one form of **hard copy**.

BLOWUP.
The enlargement of a **video image**.

BOND PAPER COPYING. *See* **electrostatic copying**.

BOOLEAN ALGEBRA.
Named for G. Boole, English mathematician and logician, a system of logic which employs algebric notation to describe logical relationships. **Boolean operators** are used to establish a variety of relationships (or associations) between and among indexing terms when they are employed in **information retrieval**.

BOOLEAN LOGIC. *See* **Boolean algebra**.

BOOLEAN OPERATOR.
In **Boolean algebra**, any of a number of logical terms which establish relationships between or among indexing terms when these are used for **information retrieval**. The most common Boolean operators are "and," "or," and "not."

BOOM, CAMERA.
A motion picture camera support by means of which the camera may be moved to different positions in order to photograph a scene from a variety of views. Camera booms are most often used for overhead shots, in which the camera looks down on the scene.

BOOM MICROPHONE.
A **microphone** that has been mounted on a boom that is a support or beam. Very small booms can be fastened to specially designed **headphone**s so the user can listen and speak to the performers who are being **monitor**ed or he can record his own voice while monitoring the recording process. Larger booms are frequently found in recording studios and are used to place the microphones in optimal recording positions.

BOUNCE (TELEVISION).
An unnatural sudden variation in the brightness (**amplitude**) of the television picture.

BRANCHING PROGRAM. *See* **programmed instruction**.

BREAK.
To interrupt, usually by means of a single key on the **keyboard** of a **terminal**, a transmission to the terminal, usually emanating from a **computer**.

BUBBLE MEMORY.
A type of **computer memory** which is composed of minute magnetizable cylinders (or "bubbles") which are at right angles to the non-magnetic sheet on which they are deposited. Each bubble represents one **bit** of memory. The principal advantage of bubble memory is its exceptionally high **packing density** when compared to most other memory forms.

46 / BUFFER

BUFFER.
A device which provides for the temporary **storage** of **data** between an **input/output** device and the **main storage,** or between any sending device and relatively slower receiving device.

BUG.
A mistake, malfunction or other such problem. Sometimes, but erroneously, synonymous with **glitch.** Bugs are usually associated with **software** such as **program**s.

BULK ERASER.
An electromagnetic device that can **erase** (i.e., remove) the recorded signals from an entire **open-reel magnetic tape,** audio **cartridge** tape or audio **cassette** tape in a matter of seconds.

BURNED-IN IMAGE (TELEVISION).
The **image** which persists in a fixed position (as an after image) in the **output signal** of a camera tube (or **pickup tube**) after the camera has been turned to a different scene. Burned-in image can occur in either the camera viewfinder or the television screen. To avoid this, the scene should not remain fixed for more than a few minutes at a time.

BURN-IN. *See* **dodging.**

BURSTER.
A machine which separates the pages of a **computer printout** very quickly, and usually all at once or sequentially.

BUTT SPLICE. *See* **splice, film.**

BYTE.
A group of adjacent **bit**s, frequently representing a **character**. A byte is usually composed of a given number of bits, ordinarily from six to eight.

BYTES-PER-INCH. *See* **BPI.**

BYTES-PER-SECOND. *See* **BPS.**

CAI. *See* **computer assisted instruction.**

CATV. *See* **community antenna television.**

CCTV (CLOSED CIRCUIT TELEVISION). *See* **educational television, closed circuit.**

CD-4.
Although no longer in production, CD-4 was the registered trade name of a **quadraphonic phonograph** disc system, sometimes called **discrete four channel** discs or quadradiscs by RCA Victor and other manufacturers. This system is different from the **matrix four channel** system of such manufacturers as Columbia and Connoisseur. The CD-4 system utilizes a regularly grooved disc into whose grooves four **channel**s of **information** have been recorded along with high frequency **carrier signal**s (ranging from twenty thousand **hertz** to forty-five thousand hertz), which must be demodulated by a CD-4 **demodulator.** Besides the demodulator, the CD-4 system requires a **phono cartridge**

CD-4 (cont'd)
and **stylus** capable of reproducing these high frequency signals, plus low-capacitance cables from the phono cartridge to the **amplifier** and such other typical quadraphonic equipment as four amplifiers and four **loudspeaker** systems.

Generally, the CD-4 approach to quadraphonic discs tends to provide a somewhat better separation of the four channel information, especially the rear left-to-right distinction, than does the matrix quadraphonic disc.

A complete **stereophonic** program results when CD-4 discs are played on regular stereophonic equipment although, obviously, the effect cannot be quadraphonic.

The CD-4 system is sometimes referred to as a "4-4-4" system.

CIJE (CURRENT INDEX OF JOURNALS IN EDUCATION). *See* **Educational Resources Information Centers.**

CMI. *See* **computer managed instruction.**

COBOL.
Acronym for Common Business-Oriented Language. A **high-level language** published first in 1960 and intended for use by the U.S. Department of Defense. As its name conveys, COBOL was devised for business applications. One of its features is that it employs understandable natural language (e.g., English) commands.

COM. *See* **computer-output microform.**

COSATI. *See* **Committee on Scientific and Technical Information.**

CPB. *See* **Corporation for Public Broadcasting.**

CPU. *See* **central processing unit.**

CrO_2. *See* **chromium dioxide tape.**

CRT. *See* **cathode-ray tube.**

CABLE RELEASE.
A device which is connected to a camera and which will release (or "fire") the **shutter** by remote control. When using a cable release, the camera is usually fixed to a **tripod.**

CABLE TELEVISION.
A means of providing good quality television **signal**s, often from remote transmitters (using either very large **antenna**s or **communications satellite**s) to subscribers who are connected to a central distribution point via a **dedicated** cable.

CALCULATOR.
Electronic device (usually powered by both **alternating current** or **direct current**, but if one only, almost always the latter) which is capable of performing at least the four basic arithmetic functions (i.e., addition, subtraction, multiplication and division), but often includes percentage calculation and square root extraction, and may include **memory** capability. Some calculators (called printing calculators) provide a paper tape record of computations, etc. Others provide only a **light-emitting-diode** or **liquid crystal display.**

CALCULATOR (cont'd)

The most expensive calculators are those which are **program**mable, that is, they will remember a given number of operations (in sequence) by means of magnetic cards or other such devices. With the large number of calculators now available, it is prudent to choose one very carefully, as the array of capabilities is great and the cost will vary accordingly.

CAMERA BOOM. *See* **boom, camera.**

CAMERA TUBE (TELEVISION). *See* **pickup tube.**

CANTILEVER. *See* **stylus.**

CAPACITOR MICROPHONE. *See* **condenser microphone.**

CAPSTAN.

A spindle, or shaft, that is sometimes termed capstan idler. In a **tape recorder**, it rotates the tape against a rubber pressure or pinch roller and pulls it through the machine at a relatively constant speed during the playback and recording **mode**s. It is disengaged in the **fast-forward** and **rewind** modes in **open-reel** and audio **cassette** machines.

Some better machines have two capstans, usually located on either side of the **head** assembly. This provides a more accurate speed for the tape.

CAPSTAN IDLER. *See* **capstan.**

CAPTIONED.

An encoding means, in television broadcasting, for including captions, or subtitles at the bottom of the picture. These captions are epitomes of what is being spoken and are intended as an aid to the hearing impaired. There are two types of captioning techniques: A) "closed captioned" which, if a special decoding device is attached to the television **receiver,** provides the otherwise normal picture with the caption, and B) "open captioned," which requires no decoder (the caption appearing on all receivers tuned to that open captioned broadcast).

CAPTURE RATIO.

The ability of a **frequency modulation** (FM) **tuner** to reject unwanted FM stations and **interference** that occur at the same **frequency** as the desired station. Capture ratio is specified in **decibel**s, and the lower the figure, the better. The capture ratio of a good FM tuner should be no higher than two decibels.

CARD PUNCH.

A mechanical device which punches rectangular holes into **Hollerith card**s as a means of encoding **data**. The card punch may be manually or automatically controlled.

CARD READER.

An electronic device which scans cards which have been encoded by a **card punch** and which logs the **data** contained thereon into some **memory** or **buffer storage**.

CARD SORTER.

An electromechanical device which arranges **Hollerith card**s (which have been encoded with **data** by a **card punch**) into one or more sequences which have been preselected (e.g., alphabetically or chronologically).

CARDIOID PATTERN.
A term applied to a **microphone** essentially designed to "hear" only those sounds that occur in front of it. Good cardioid microphones are most receptive to sounds occurring within 180 degrees of a line drawn at right angles to the length of the microphone and through the front of it. They are least receptive to sounds emanating to the rear of that line. The other basic type of response pattern in microphones is found in the **omni-directional microphone**.

CARREL.
A carrel was originally a small space or area in the stack room of a library which was set aside for students and patrons to use for studying and reading. Today, study carrels are desk-like devices that are produced by many manufacturers and that come in a wide variety of sizes, styles, and arrangements. They may be the central focus of a library, learning resources center, or **media center**. A "wet" carrel is one that is provided with electrical or electronic services, whereas a "dry" carrel consists of the furniture only. Depending on the number of electrical outlets, a wet carrel may use one or more pieces of audio-visual equipment (such as a tape player or a **filmstrip** projector) and may provide for **rear screen projection**. It may also have **terminal** outlets for television and for **audio** program reception, and connecting lines to **computers** or other **data** services.

CARRIER.
A wave, usually radio, of known and constant **frequency, amplitude,** and **phase**, which can be varied by changing its amplitude, frequency, or phase. In broadcasting it is used to "carry" the **modulation** information in both **amplitude modulation** (AM) and **frequency modulation** (FM) broadcasts. The term *carrier* is also used to describe the rather high frequency information in **CD-4 quadraphonic discs**, which enables the **demodulator** in this system to sort out the **audio** portion of the **signals** specific to each **channel**.

CARTRIDGE, AUDIO.
A type of enclosed **magnetic recording** tape which is distinct from the **open-reel** format. Enclosed within a plastic case (6x4x¾-inches) is an endless loop of ¼-inch width **magnetic tape** which travels at 3¾ **inches-per-second** when inserted into a tape cartridge player or recorder. The tape in an audio cartridge plays continuously, since, as stated above, it is formed of an endless loop. One disadvantage of the audio cartridge system is that it cannot be rewound to a specific point. Usually, four separate **stereophonic** programs (i.e., eight **tracks**) are recorded on the tape.

CARTRIDGE, PHONOGRAPH. *See* **phono cartridge**.

CARTRIDGE PROJECTOR. *See* **film cartridge**.

CASSETTE, AUDIO.
A **magnetic recording** system similar to the **open-reel** format, but much reduced in size and enclosed in a small plastic container (4x2½x½-inches). Devised by the Philips Lamp company of Holland, it uses two small **reels**, and each of the ends of the **magnetic tape** is attached to a separate reel. The tape itself is almost one-eighth inch wide; in the audio cassette recorder or player it travels at 1⅞ **inches-per-second**. At the rear of the cassette are two small tabs, either or both of which may be broken off to prevent accidental re-recording of one or both sides of the tape. At present, only **monaural** and **stereophonic** cassette players are commercially available. Unlike many open-reel recorders and players, cassette devices are monaurally and stereophonically compatible; that is, either type of

CASSETTE, AUDIO (cont'd)

recording may be played (with no loss of recorded material) on either a monaural or a stereophonic cassette device.

Many of the better audio cassette recorders and players now come equipped with the **Dolby noise reduction** system and are able to accept, by means of switch settings, any of the currently available tape formulations.

Because of various factors, audio cassette recorders and players are usually limited in **frequency response** from about thirty **hertz** to eighteen thousand hertz plus or minus three **decibel**s.

Audio cassettes come in various playing time lengths, the most common of which are 30, 40, 60, 90, 120, and 180 minutes (playing time is the total time for both sides).

CASSETTE DECK. See **deck**.

CATHODE-RAY TUBE (CRT).

A vacuum tube inside of which is a focused beam of electrons aimed at a portion of the front of the tube called the **raster**. Such tubes are used in television sets (and therein are called picture tubes); **computers**, oscilloscopes; in miniature form, as tuning indicators in **receiver**s (where they are usually called **magic eye**s); and as **monitor**s when computerized **information** is to be shown.

CENTERING METER.

A device used in a **receiver** or **tuner**; when a station is precisely tuned, the indicator needle in the centering meter is centered between two markers.

CENTRAL PROCESSING UNIT (CPU).

The portion of a **computer** which includes **memory**, **control unit** and **arithmetic unit**. It is the central processing unit which carries out the manipulation of **data** essential to the concept of the computer.

CERAMIC CARTRIDGE. See **piezoelectric phono cartridge**.

CHALKBOARD.

The chalkboard or blackboard is one of the oldest audiovisual devices used in teaching. Originally, chalk was used for writing on large slabs of slate. Most modern boards, however, consist of heavy cardbound material that has been stained or painted black (or, in some cases, green). Certain modern blackboards cannot be cleaned by using water and a wet rag—only a special board cleaner (similar to a large eraser) may be used to brush off the excess chalk.

A modification of the basic chalkboard is the magnetic chalkboard, which has a backing made of steel. Lettering, sketches, diagrams, or photographs can be mounted on cardboard, with magnets attached to the back, and can be displayed on the magnetic chalkboard.

All sorts of diagrams, outlines, maps, etc., may be very neatly drawn on a chalkboard by using templates that can be purchased from most professional school supply sources. Also, the **overhead projector** can project transparencies with simple diagrams or outlines onto the chalkboard, where they can be traced. Still another technique is to design one's own **pounce pattern** drawings for chalkboard display.

CHANGER, RECORD. *See* **turntable**.

CHANNEL.
An isolated **signal** path. A **stereophonic** system has two channels, designated "left" and "right." **Monaural** systems have but one channel, and **quadraphonic** systems have four: left front, right front, left rear, and right rear. Occasionally, the term *channel* is confused with **track**.

The term *channel* can also be used to mean the **frequency band** assigned to a radio or television station for broadcasting purposes.

CHANNEL SEPARATION.
A specification indicating the degree to which two or more **channel**s are isolated from each other in **stereophonic** and **quadraphonic** systems. The specification is always given for a **frequency range** in **decibel**s, and the higher the number the better.

Channel separation for a **frequency modulation** (FM) **tuner** should be at least thirty decibels for frequencies from thirty **hertz** to fifteen thousand hertz. As a matter of fact, twenty-five to thirty decibels of channel separation is adequate for a good stereophonic effect in most equipment.

CHARACTER.
The symbolic representation of numbers, letters or other such. A character may be **code**d (e.g., for the purpose of a **computer**) or representational (as are the typed or printed letters of the alphabet).

CHARACTER SET.
A unique and complete ensemble of **character**s which make up a distinct group recognizable to a **computer**. Examples are the Cyrillic alphabet; the Morse **code** alphabet; **Boolean operator**s; etc.

CHECK DIGIT.
A digit conveying no intrinsic **data**, but included with a group of meaningful data in order to enable the **computer** to discover errors in a number of possible stages.

CHIP.
A small, individual piece of a silicon crystal which contains the equivalent of multitudes of **solid-state** electronic devices. A chip can range in size from ½-inch square to the virtually microscopic. Chips are used in a **computer** and form the principal electronic portions of the **central processing unit**.

Chips are the devices on which **integrated circuits** (be they large-scale integrated circuits, LSI, or very large-scale integrated circuits, VLSI) are formed.

CHROMA.
A technical term that pertains to saturation or the amount of hue in color. There is, for example, a great deal of red in vermilion, but not much red in pink. Sometimes hue is equated to color, but it is specifically used to indicate a modification of a basic color, e.g., orange of a reddish hue.

CHROMATIC ABERRATION. *See* **aberration**.

CHROMIUM DIOXIDE TAPE.

A **magnetic recording** tape whose **coating** (i.e., the portion of the tape on which the actual recording is made) is a newer formulation than ferric oxide, which has been in use as a coating for decades. Chromium dioxide (abbreviated CrO_2) tape has a higher magnetic coercivity than **ferric oxide tape**. This means that CrO_2 tape requires more demagnetizing force because its magnetic state is more resistant to change. This characteristic is important in CrO_2 tape because it is able to record and maintain sounds of higher **frequency**. The importance of this feature is evident when the **tape recorder** employed operates at slower **tape speed**s, as in audio **cassette** systems (1 ⅞ **inches-per-second**) or audio **cartridge** systems (3¾ inches-per-second). The reason for this is that the slower the tape speed the more difficult it is to record high frequencies. In order for a tape recorder to use CrO_2 tape, it must have the proper **bias** frequency. Newer tape recorders tend to have a bias frequency switch for both CrO_2 and ferric oxide (properly Fe_2O_3, but more often, simply FeO) tapes. Chromium dioxide tape is **tape type II**.

The use of CrO_2 tape may effect a rather dramatic change in the **frequency response** of a tape recorder. For instance, a cassette tape recorder using FeO tape may have a frequency response of thirty **hertz** to ten thousand hertz plus or minus four **decibels**. However, with the use of CrO_2 tape, the frequency response of the same machine may well be thirty hertz to twelve thousand or even fourteen thousand hertz plus or minus three or four decibels.

The only real disadvantages in the use of CrO_2 tape to date have been that it is relatively abrasive (so that it tends to wear out the **heads** of the tape recorder somewhat sooner than does FeO tape) and that it does not record **bass** frequencies as well as FeO tape does. A newer formulation (ferrichrome) does offer the desirable frequency response of both CrO_2 and FeO tapes, with less abrasiveness. But both CrO_2 and **ferrichrome tapes** (tape types II and III, respectively) have, since about 1980, been supplanted by **metal tape**.

CINCH MARKS.

A series of straight scratches, varying in length, along the length of a **reel** of **film** and caused by pulling the film too tightly when winding the film onto the reel.

CINÉMA VÉRITÉ.

A type of **documentary film** that has been produced with mobile (or hand-held) camera and sound equipment. This enables the film-maker to get closer to his subjects and thus make a film that has more "veracity."

CLAW MECHANISM. See **pawl-sprocket**.

CLEAN.

In **audio** or **video**, refers to **distortion**-free reproduction.

CLIPPING.

The deformation of a wave form which is ordinarily of a sinusoidal shape. This deformation usually occurs when an **amplifier**'s **power output** has been so increased (in **amplitude**) that the device can no longer reproduce the actual shape of the **input** wave form. A wave form that looks rather like an S on its side and that has the two curved portions flattened, or clipped, is a form of **distortion**. Clipping may also occur in other devices (e.g., the **preamplifier**, the **tuner**), but it is most common in amplifiers.

CLOSE-UP.

A shot in which the camera is, or appears to be, very close to the subject. A close-up of a person includes only the head, or the head and shoulders.

CLOSED CAPTIONED. See **captioned**.

CLOSED CIRCUIT TELEVISION. See **educational television, closed-circuit**.

CLOTHBOARD.
Refers to a number of different visual display devices which are called by the name of the materials from which they are made. These include flannelboard, feltboard, flockboard, Velcroboard, and Hook 'n Loop Board. Convenient and inexpensive clothboards can be made from flannel because the pieces stick together when gentle pressure is applied. Letters, simple shapes, or outlines can be cut out and applied to a piece of flannel stretched over a sheet of plywood or particle board. Also, pieces of flannel can be attached to pictures or pieces of cardboard and placed on the flannelboard. Felt, although it is more expensive, can be used in the same way. However, because these materials have a tendency to slip off the board, better materials have been marketed. Velcro, which is made from nylon, is an example of a more effective material. Its surface consists of many tiny nylon hooks, which interlock when two pieces are pressed together. Hook 'n Loop is a trade name for a Velcro type material.

COATED LENS.
A lens which has had a thin, transparent (and relatively hard but scratchable) coating applied to its surfaces in order to reduce reflection within the lens and to improve transmission of the **image**. Care should be exercised when handling or cleaning coated lenses.

COATED PAPER COPYING. See **electrofax copying**.

COATING.
The magnetizable material used in **magnetic tape** for **magnetic recording**. The coating, which is applied to the **base**, is of one of four formulations, and results in designation of the tape to which it is applied as one of four **tape types**: **ferric oxide** (tape type I) is the oldest and still the most commonly used; **chromium dioxide** (tape type II) which, like **ferrichrome** (tape type III), has fallen into virtual disuse; and **metal tape** (tape type IV), which has become popular since about 1980.

COAXIAL CABLE.
An insulated cable that has, starting at the cross sectional center, a conductor, insulation (usually of light plastic or paper), a braided metal shield to prevent the **interference** of other fields (either electrical or magnetic), and outer insulation. Coaxial cables may have more than one conductor, and the shielding is ordinarily used to form the grounding lead (the inner conductors sometimes are designated as the **signal** leads).
This type of cable is frequently used for **audio** and television connections, and it is almost always used as the better television **antenna lead-in**.

COAXIAL SPEAKER.
A **loudspeaker** consisting of a **woofer** with a **tweeter** or other such unit mounted centrally, usually in front.

CODE.
The means by which **data** are represented in some alternate symbolic form which is recognizable (and capable of being manipulated) by a **computer**. The **binary** system is one such code.

54 / COLLAGE

COLLAGE.
A type of art work in which all sorts of heterogeneous objects and materials are pasted together in an incongruous relationship for their symbolic or suggestive effect. This term is also applied to a flat picture, photograph, or bulletin board display in which disparate figures, shapes, pictures, parts of pictures, etc., are pasted together in an oddly shaped array.

COLLATOR.
A mechanical device, usually attached to a **copying machine**, which automatically places finished copies into ordered sets.

COLOR NEGATIVE FILM.
A type of color **film** whose **negative**'s colors are complementary. Color negative film is usually used to produce color **prints** but black and white prints or color transparencies can also be made from it.

COLOR POSITIVE FILM. *See* **color reversal film**.

COLOR REVERSAL FILM.
A color **film** (commonly called **slide** film) whose **negative** is actually a **positive** (i.e., after processing, the film's colors are normal: blue is blue, red is red, etc.) **transparency** or slide. Compare with **color negative film**.

COLUMBIA FOUR CHANNEL. *See* **matrix four channel**.

COMMITTEE ON SCIENTIFIC AND TECHNICAL INFORMATION (COSATI).
A standardization group made up of **information** executives (active from the early 1960s through the early 1970s) which set, among other things, a standard for **microfiche** format (4x6-inch sheet containing sixty **frames** in five rows and twelve columns, each **image** having a **reduction ratio** of eighteen to one). Although not truly binding (e.g., with the aforementioned microfiche format), many of these standards are still observed.

COMMON CARRIER.
A public utility, or private corporation controlled by a public (and legally constituted) administrative agency, which conveys either **data**, **information**, or tangible property. An example of such a utility is the American Telephone and Telegraph Company. Other such private common carriers are moving and storage companies and long distance telephone companies like MCI and **Telenet**.

COMMUNICATION LINK.
That equipment which is required to connect one location to another in order to exchange **data** by transmission.

COMMUNICATIONS COMMON CARRIER. *See* **common carrier**.

COMMUNICATIONS SATELLITE.
A space satellite (usually stationed 23,500 miles above earth and orbiting in such a way as to appear fixed over a given geographic locus) which can receive **data** or **information** from one location on earth and transmit it to another. It can be, and is, used for the transmission of telephone conversations, television broadcasts, **computer** data, etc.

COMMUNITY ANTENNA TELEVISION (CATV).

A redistribution system, now in widespread use, that receives television **programs** from regular broadcasting stations by means of a common **antenna** and then relays them via an **educational television, closed circuit** system to cable service subscribers in a particular area. CATV systems initiate television programs either on **magnetic tape** or **film** or live for local viewing. CATV is characterized by high quality and reliable reception, but, most importantly, cable service can transmit as many as thirty separate programs simultaneously. It was originally hoped that as much as 20 percent of cable capacity would be devoted to educational purposes, but as of 1972 the **Federal Communications Commission** has allotted only one free **channel** to education.

COMPATIBILITY.

This term has various senses in audiovisual discussion. It can mean any of the following:

1. A color television system that can also receive and display black-and-white pictures.
2. A **stereophonic** system that can also reproduce **monaural signals**.
3. A **quadraphonic** system that can also reproduce stereophonic or monaural signals.
4. A stereophonic **frequency modulation** (FM) broadcasting system whose signals can also be received as monaural.
5. An 8mm **film** system capable of using regular 8mm or Super 8mm film.

Generally, the term is taken to mean equipment or systems that are not mutually exclusive.

COMPILE.

The act of creating a **machine language program** from a **computer** program written in another **programming language** by using a special program called a **compiler**.

COMPILER.

A **computer program** which translates another computer program from a **high-level language** into **machine language**.

COMPLEMENTARY COLOR.

Opposite or contrasting color. **Filters** which pass a given color "subtract" (or block) colors complementary to those which pass through the filter.

COMPLIANCE.

The flexibility of the **stylus** in a **phono cartridge**. This refers particularly to the ease with which the stylus can follow the changes in side-to-side direction within the grooves of the **phonograph disc**. Since compliance is determined intentionally by manufacturers' designs, a more useful specification is the minimum recommended **tracking force**.

The term *compliance* is occasionally used with respect to **loudspeakers**. High compliance speaker is a meaningless description; it is used in the hope of creating in the potential purchaser's mind the idea of a very flexible loudspeaker **cone**.

COMPUTER.

An electronic device (usually with some mechanical parts) which can, by means of rapid computations, calculations, comparisons, etc., provide solutions to problems—all without human intervention, once it is properly **program**med and is fed the appropriate **data**. Contrary to some popular notions, computers neither "think" nor "feel" as humans can.

COMPUTER (cont'd)

Computers are of two basic types: **analog** and **digital**. The analog computer represents, measures, and manipulates discrete data (such as individual numbers) by means of such physical variables as voltage changes or step-down in mechanical gears. Similar to such analogs are slide rules, speedometers, etc. Notice that representations within the limits of such devices are within an infinity of possibilities. Digital computers, on the other hand, like the **calculator,** adding machine, and abacus, deal with discrete data as such on a "quantized" basis. One of the basic differences between analog and digital computers is that the analog can manipulate and represent data directly, while the digital must have the data transformed into some **code**.

Almost all computers are divided into smaller subsystems: **memory, arithmetic unit, control unit, input/output,** etc.

There are three size classifications of computers based on memory and computational capability: **microcomputer**s, **minicomputer**s, and **main frame** computers.

COMPUTER ASSISTED INSTRUCTION (CAI).

A system of individualized instruction which involves direct interaction between learner and **computer**, with minimal intervention from a human instructor. Computers can individualize instruction, and they can provide valuable **data** on how students learn (see **computer managed instruction**). If a school district cannot afford to buy a computer, it can lease time on one; then, by means of **terminal**s located at the schools, students can have access to many different **data banks**.

The computer is an extremely versatile device. It can display pictures and diagrams on a **cathode-ray tube** by using **film**s or **slide**s. In some cases it can show three-dimensional views, which can be rotated through different planes by the operator. It can use a **tape recorder** to give spoken instructions, and some computers have been "taught to talk." A student can respond to the **image** on the cathode-ray tube by putting a **light pen** at specific places, and his answers will be evaluated. Since a computer has both **memory** and "logic," it can remember all the responses of the student, classify them, and sum them up. The student can use the computer to solve problems, to practice learning skills, to ask and get answers to questions, or to learn from simulated situations.

The computer is perhaps the best teaching machine that has ever been designed for **programmed instruction**. However, the effectiveness of a computer-taught instructional program depends on the quality of the program (most succinctly expressed in the epigram GIGO: **garbage-in-garbage-out**). Idiomorphic programming is representative of an attempt to maximize the use of the computer's flexibility in order to individualize instruction. It is a special type of tutorial instruction "in which decisions are made regarding the next instructional experience of the student in terms of entry behavior data and sets of responses made during learning. In idiomorphic programming, the repertory of the teaching system is more extensive than it is when teaching is conducted using other models. The four types of options (repeat, remediate, skip, or pace) are decided upon in terms of a larger base of **information** about the student than is used in other branching programs such as intrinsic programming" (Lawrence M. Stolurow, *The Schools and the Challenge of Innovation*, CED Supplementary Paper 28 [New York: Committee for Economic Development, 1969], 317-18).

COMPUTER LIBRARY. *See* **library, computer.**

COMPUTER MANAGED INSTRUCTION (CMI).

Modern electronic **computer**s have been used in education since the early 1960s. Depending upon costs, they may play an ever-increasing role in direct instruction

COMPUTER MANAGED INSTRUCTION (CMI) (cont'd)

(computer assisted instruction) and in evaluation of student progress. CMI is a system of educational management that uses the computer to help the teacher individualize instruction by providing up-to-date **information** about student backgrounds, interests, achievements, and needs.

COMPUTER-OUTPUT MICROFORM (COM).

A **microform** produced from **data** which a **computer** has manipulated and then displayed in such a way as to create the microform without first producing **hard copy**.

COMPUTER PROGRAM. *See* **program**.

COMPUTER WORD.

A sequence of **characters** which are treated as a unit and are stored in one **computer memory** location.

CONDENSER MICROPHONE.

A **microphone** whose operation depends on changes in its electrostatic capacitance. The condenser microphone differs from the **dynamic microphone** in that it requires its own supply of electrical power. The condenser microphone also differs from the dynamic microphone because its **frequency response** and **dynamic range sensitivity** are usually better.

A new version of the condenser microphone is the "electret" condenser in which the power supply, ordinarily a single AA penlite battery, can be housed within the microphone itself. Condenser microphones, because of their comparative complexity, tend to be less rugged than dynamic microphones, yet they are to be preferred for better sound quality.

Two of the most important specifications for microphones are **frequency response** and **polar pattern**.

CONE.

This is the moving portion of a **loudspeaker** (of the **piston speaker** type); when activated by electromagnetic energy, it translates that energy into sound waves in the air.

CONICAL STYLUS.

A **phonograph stylus** whose cross section is circular and whose tip is rounded. The conical stylus, sometimes called the spherical stylus, is most useful in the reproduction of 78 rpm or other **monaural discs**. Often found in less expensive **phono cartridge**s, it is not adequate for the best reproduction of **stereophonic** discs (for which an **elliptical stylus** is good).

CONNECT TIME.

The amount of time a user is **online** to a **computer** (often used when the computer is remote to the user and the connection is made via telephone line) regardless of whether the computer has been operating or idling.

CONSTANT MEMORY.

In a **calculator**, a **memory** which retains the values entered into it even when the calculator is switched off. It does this by feeding a minute electrical current to the unit from its power supply (similar to the clock in an automobile) — this current is frequently called a trickle drain. Constant memory is often referred to as non-**volatile memory** or permanent memory.

58 / CONSTRUCTED RESPONSE PROGRAM

CONSTRUCTED RESPONSE PROGRAM. *See* **programmed instruction**.

CONTACT PRINT.
A photographic positive **print** usually made for test purposes, created by placing the **negative** in physical contact with the print paper, exposing them to light, and then chemically processing the print paper. Some contact printing, which is done as inexpensively as possible, creates a print which is unstable and whose **image** fades after several weeks. This type of print is often referred to as a test print or proof. It is used to determine which negative should be selected for full **enlarger** treatment and processing (processes which are rather more costly).

CONTINUOUS LOOP FILMS. *See* **film**.

CONTINUOUS POWER.
A specification related to the **power output** of an **amplifier** (sometimes referred to as root-mean-square [RMS] power). It is obtained by feeding the amplifier a continuous, usually pure, tone and measuring the amplifier's ability to increase the **amplitude** of that tone before being driven into **distortion** (usually **clipping** or **harmonic distortion**). Various methods have been used to measure amplifier power output, but continuous power is generally regarded as the most accurate and reliable method. The Federal Trade Commission now requires manufacturers to specify output power in terms of continuous power.

CONTINUOUS PRINTING.
A method of printing whereby the **film** to be copied and the light sensitive paper onto which the **image**s will be copied are synchronized in their movement in order for the **print**s to flow continuously from the device into which they are fed.

CONTRAST.
A range of light and dark values in a picture, or the ratio between the maximum and minimum brightness values. For example, a high contrast television picture has intense blacks and whites whereas a low contrast picture contains only various shades of grey.

CONTROL UNIT.
That portion of a **computer** which supervises the functioning of the computer by calling upon other units as they are required by specific **program**s. It is one portion of the computer's **central processing unit**.

CONVENIENCE OUTLET.
A **jack** on a device (e.g., **amplifier, motion picture projector**) which accepts the **line cord plug** of another device.

COOK SYSTEM.
An early (circa 1950) **stereophonic** type **phonograph disc** recording system that simultaneously cut two grooves into the **disc**, one for each stereophonic **channel**. Playback was obtained by having two **phono cartridge**s in the **tone arm**, placed parallel to each other and simultaneously tracking the concentric grooves. The Cook System is sometimes called a binaural phonograph.

COPYING MACHINE.
Any device which copies the **image** on a surface onto a blank surface, usually taken to mean a machine which copies printed material onto blank paper. There are two systems of such copying now commonly in use: **electrostatic copying** and **thermographic copying**, the former being the most common.

CORE.
A type of **computer memory** common from the late 1950s through the early 1970s. A core is a small (approximately one-sixteenth of an inch in diameter) toroidal- (or donut-) shaped piece of magnetic material. Each core is capable of retaining one **bit** of **data**. Many cores are strung together with electrical current-conducting wires (rather like the covering on a beaded handbag) which can magnetize or demagnetize individual cores. Because of the cost of manufacturing and "stringing" cores, this type of memory has given way to other, cheaper ones, like **bubble memory** and memory on **chip**s.

CORPORATION FOR PUBLIC BROADCASTING (CPB).
A nonprofit, nongovernmental corporation established by Congress in 1967 to promote the growth and development of noncommercial radio and television. It can receive funds from both public and private sources, and these are used to support local **public television** and radio services. It serves to coordinate national radio and TV services by distributing programs, increasing the number and quality of (and inventorying) programs available for use by local stations, and supporting the **Public Broadcasting Service** (PBS) generally by means of talent development, innovational techniques, audience research, and public **information**. The CPB has been instrumental in setting up the national network of public television through its various agencies and affiliated organizations such as the Public Broadcasting Service, **National Educational Television** (NET), and **The National Association of Educational Broadcasters**. Address and phone number for the CPB: 1111 Sixteenth Street, N.W., Washington, D.C. 20036, (202) 293-6160.

CORRECTIVE MAINTENANCE. *See* **maintenance**.

COUNTER, INDEX. *See* **digital counter**.

CRASH.
A **computer** failure, either of **hardware** or **software**, which causes the entire system to be non-functioning. One serious type of crash is the **head crash**.

CROPPING.
In photography, the process of trimming off unwanted portions of a picture. Cropping can be done by physical excision or, in the dark room, by means of an **enlarger**.

CROSSOVER.
A device in a **loudspeaker** system that divides the **frequency range** into **band**s and feeds each band to the specific loudspeaker appropriate for that band. For example, a band of frequencies from 20 **hertz** to 150 hertz might be fed to the **woofer**. The crossover **frequency** comprises the center of the frequency limits of each band, and separates them. If, for example, the first band is 20 hertz to 140 hertz and the second band from 160 hertz to 3,000 hertz, then the crossover frequency is 150 hertz.

CROSSOVER FREQUENCY. *See* **crossover**.

CROSSTALK.
A specification of the undesirable leakage of a **signal** from one **channel** to another; for example, right channel **information** may leak into the left channel. This specification is the opposite of **channel separation**.

The specification is given as a minus number in **decibels**. It indicates, in decibels, how far the unwanted signal is below the desired signal. A minimum crosstalk specification for **stereophonic** systems should be about minus thirty decibels. The greater the minus number, the better (minus fifty decibels crosstalk is better than minus thirty decibels).

Crosstalk, as a specification, can be applied to **phono cartridge**s, **tuner**s, **amplifier**s, **receiver**s, and **tape recorder**s.

CRYSTAL CARTRIDGE. *See* **piezoelectric phono cartridge**.

CUEING.
The ability to select a specific point on a **phonograph disc** or **magnetic tape**. Cueing allows one to select a specific song on a given disc and place the **stylus** at the beginning (or any other portion) of that song. Phonographs that allow for cueing usually have some sort of mechanical device to lift the **tone arm** above the surface of the disc, allowing the user to move the tone arm to the desired spot and then lower it. **Open-reel** type **tape recorders** usually accomplish this by means of a **pause control**, which keeps the tape in contact with the **head**s; then, while the motor drive system is disengaged from the tape (causing the tape to stop while the drive system idles in a form of "neutral"), one rocks the **reel**s back and forth to a specific point, either for **editing** purposes or simply to begin playing from that point. **Automatic search** is another type of cueing.

CURRENT INDEX OF JOURNALS IN EDUCATION (CIJE). *See* **Educational Resources Information Centers**.

CURSOR.
A **character** on a **computer cathode-ray tube** (usually a rectangle whose long side is vertical, or an underscore one character in width) which denotes the place of the next character which will be generated. Frequently the cursor blinks on and off to call attention to itself and to prevent a **burned-in image**. The placement of the cursor is controlled by a set of cursor control keys on the **terminal keyboard**.

CYCLES-PER-SECOND. *See* **hertz**.

DA. *See* **direct access storage**.

DAS. *See* **direct access storage**.

DASD. *See* **direct access storage**.

dB. *See* **decibel**.

DBMS. *See* **database management system**.

DBS. *See* **direct broadcast satellite**.

dbx.
A system, similar in intent and effect to the **Dolby noise reduction** one, but considered to be more efficacious in reducing **tape hiss**. It is named for the manufacturer which produces it, and is always written in lower-case **characters**.

D/C. *See* **direct current**.

D/C RESTORATION.
The ability of a television picture tube to display changes in brightness as the television camera "sees" them. Good D/C restoration is most noticeable when large portions of the picture are very dark, as in scenes at night, and objects (actors, etc.) are moving about.

DF. *See* **damping factor**.

DOS. *See* **disk operating system**.

DAISY WHEEL.
A type of **printer** in which the printing element has a plastic or metal hub with radial spokes emanating from it like the petals of a daisy. On each spoke there is a **character**. The hub rotates and stops rapidly, and when a desired character is in the proper position, the character is struck by a hammer which drives it against an inked ribbon, and thereby creates a printed impression on the paper behind the ribbon. This is one type of **impact printer**.

DAMPING.
Applied to devices such as **loudspeakers**, this term refers to the device's ability to reduce undesirable (and residual) loudspeaker **cone** movement. Other devices may also be spoken of as damped—e.g., the **stylus** of a **phono cartridge** can be damped to prevent high **frequency distortion**. The term *damping* originated with orchestral cymbals players who would wet their fingers prior to pinching a vibrating cymbal to stop its sound.

DAMPING FACTOR (DF).
A specification applied to **audio**-style **amplifiers** which denotes the amplifier's ability to reduce (dampen) unwanted, residual **loudspeaker cone** motion. It is usually expressed as a ratio between the amplifier's **output** impedance and the loudspeaker's **impedance** and is given as a discrete number (e.g., DF 15). The higher the damping factor, the better, with a minimum of at least DF 10.

DATA.
A general term usually meant to describe letters, numbers, and other symbols which, of themselves, do not necessarily convey **information**. It stands, somewhat, in relation to information as information does to wisdom. It is on data that the **computer** performs its operations, and it is in this sense that it is ordinarily used.

DATA BANK.
Although frequently used as a synonym for database, a data bank should be thought of as a kind of collection of "libraries" of **data**. It is these libraries which are, in fact, the databases. An example might be the Lockheed Dialog data bank which includes such databases as those of the National Technical Information Service, the **Library and Information Science Abstracts (LISA)**.

DATABASE.
A collection of related **data** or **information**, often bibliographical, but not necessarily so. A collection of databases is a **data bank**.

DATABASE MANAGEMENT SYSTEM (DBMS).
A piece or set of **software**, it is the means for handling the **data** in the **database**. In addition to this function, the DBMS must provide integrity (i.e., protection against **hardware** or software malfunctions), data independence (obviating or minimizing reprocessing or, even, reprogramming when there are changes made to the database), and privacy (i.e., prevention of unauthorized use or alteration of the data or system).

DATA COMPRESSION.
A means of saving **storage** space achievable by **code** and by elimination of redundancies, empty (or partially empty) **field**s, etc. For example, a blank line of text can be represented by a "blank line" **character**, rather than by a "blank space" character for each space in the line.

DATA FILE.
A set of **data** records, usually closely related in subject, and organized in a specific manner. Frequently, a data file is used by one person or, if by more people, by a group whose needs are closely shared (e.g., "a payroll data file," "a circulation records data file").

DATA-PHONE.
A trademark of the American Telephone and Telegraph Company for a device which allows **digital data** to be transmitted over telephone lines at a variety of speeds.

DATA PROCESSING.
This, for the most part, is what a **computer** "does." That is to say, it is the operating, combining, manipulating, etc., of or upon **data** by an electronic means in order to produce **information** or other data which is more usable or comprehensible to human requirements. Data processing is frequently referred to as ADP, the abbreviation for "automatic data processing."

DATA SET.
1. A device which acts as the link between a **computer** and a telephone line. Synonymous with **modem**.
2. A collection of **data**, the members of which are related but not necessarily composed of all the same elements. For example, a library catalog's call numbers may not all include a work or edition number; some personal addresses in a directory may not contain an apartment number.

DEBUG.
The act of finding and removing errors, faults, or problems in a **program** or **computer**.

DECIBEL (always abbreviated dB).
The standard unit of the relative intensity (**amplitude**) of either sound or an electrical **signal**. It is a logarithmic unit; 1 decibel is ostensibly the softest sound audible to the average human ear. Most humans cannot notice changes in the amplitude of sound that are less than 1 or 2 decibels. Being logarithmic, the decibel is indicative of a ratio between two amounts of **loudness** (or of electrical power). Twenty decibels are actually one hundred times greater than 10 decibels, not two times greater. The difference between 10 decibels (few homes or offices are that quiet) and 110 decibels (approaching the loudest passages in classical orchestral music) is one trillion times, not eleven. Sustained sound amplitude in

DECIBEL (cont'd)

excess of 120 decibels can be painful, and above 130 decibels may cause temporary or permanent hearing loss. The noise at rock concerts, by the way, has often been measured in excess of 120 decibels (and even 130 decibels).

The decibel is frequently used in specifications for **audio** equipment, and it is an important evaluation factor. To understand the function of the decibel in these instances, look up the pertinent specification in this dictionary.

DECK.

1. A **magnetic recording** device consisting of a **tape transport** and possibly **recording amplifiers**, and perhaps **output preamplifiers**, meters of various sorts, and operating controls; however, it lacks **output amplifiers**, **loudspeakers**, and/or a **cathode-ray tube monitor** (this latter in **videotape** or **video cassette** decks).

Some decks, usually designated as **playback deck**s, have no recording capability at all. Their advantages are that they cost less and that the user cannot erase a program. Occasionally this type of deck is used as a device to copy a program from one **magnetic tape** recording to another. The tape to be copied is played on the playback deck.

2. In **computer** parlance, a deck is a set or stack of **punched card**s.

DECODER.

In **audio** terminology, this was first used to describe the circuit in **multiplex** type **frequency modulation (FM)** broadcasting that separated the left **channel signal**s from the right channel signals. At the present time, it has also come to mean a device that sorts out each signal in a **matrix four channel** type **phonograph disc** and assigns each signal to its proper channel.

DEDICATED.

Any communication system, **hardware**, or such like, set aside for specific use and not intended for any other; e.g., a "dedicated line" being a communications line intended, perhaps, only for certain (and predetermined) **data** transmission. A dedicated line is usually in contrast to a **dial-up** line.

DEFINITION.

In **video** reproduction, the apparent sharpness and clarity of the **image** reproduced, be it a photograph or a television picture. It is essentially a subjective measure as compared to **resolving power**.

DEGAUSS.

The process of removing, by means of a degausser or demagnetizer, the unwanted buildup of magnetism on **magnetic recording head**s. This is accomplished by waving the degausser in a circle just over (i.e., approximately one-eighth inch above) the head **gap**s, for about ten seconds, being careful not to touch the heads. If the magnetism on the heads is allowed to build up indefinitely, permanent damage to the heads (and to **magnetic tape**s that come in contact with those heads) can result. Therefore, degaussing should be done after about every one hundred hours of use.

Some newer **tape recorder**s have circuits that provide automatic degaussing.

DELIMITER.

A special **character**, or group of characters, which separate groups of other characters (or **data**).

DEMAGNETIZE. *See* **degauss**.

DEMODULATOR.
1. A device used in **CD-4** type **quadraphonic phonograph** discs which sorts out the four **channel** information from the **signal**s fed to it by the **stylus**, and assigns these signals to their proper places in the quadraphonic system, i.e., to the **amplifier** and thence to the **loudspeaker**s.
2. A device which retrieves **data** from a **carrier** wave which has been processed by **modulation**. The demodulator makes the data compatible with **data processing** equipment.

DEPTH OF FIELD.
This is a measure of the focus of a camera at a given **aperture** setting. It is obtained by extending an imaginary line from (and at a right angle to) the **film plane** into the scene being photographed or **videotape**d. The depth of field would be that distance along the imaginary line which includes the closest point in acceptable focus to the farthest point in rather sharp focus. Since depth of field is releated to the camera's aperture setting, the smaller the aperture used, the greater the depth of field.

DENSITY. *See* **packing density**.

DERIVED FOUR CHANNEL.
A method of obtaining artificially synthesized **quadraphonic** sound from ordinary, **stereophonic** program sources. Many **matrix four channel** type **decoder**s have the circuits to perform this function, given a sound system that includes a sufficient number of both **amplifier**s and **loudspeaker**s. Derived four channel is occasionally referred to as a "2-2-4" system.

DEVELOP.
In photography, the process of transforming a **latent image** to a visible image. This usually involves placing the **negative** in a chemical solution which removes a portion of the silver **emulsion** on the **film**. There are additional steps necessary to produce fully **processed film**. **Vesicular film**, however, requires but heat and light to develop it.

DIAL-UP.
The use of an ordinary telephone to establish a communication link, via a telephone line, between a **terminal** and a **computer**. By means of a dial-up connection, a **dedicated** line is obviated. Dial-up usually entails a **switched line**.

DIAMETER (PHOTOGRAPHY).
A measure of linear **reduction** or **enlargement**. For example, a reduction of twenty-four diameters means that the object has been reduced linearly twenty-four times (also expressed as a reduction of 24x) and occupies an area 24^2 less (or 1/576 the space). Magnification range is another way of expressing diameter.

DIAPHRAGM.
The moving portion of a **microphone** (or an **electrostatic headphone** or **electrostatic speaker**), which is activated by (or causes) sound waves in the air. Many diaphragms now are made of a tough plastic, like mylar.

DIAPHRAGM SHUTTER. *See* **between-the-lens shutter**.

DIAZO FILM.
A polyester **film** whose photosensitive surface is one containing diazonium salts. On **exposure** to light concentrated in the blue-through-ultraviolet portion of the spectrum, a **latent image** is formed. In order to transmute this to **processed film** the process incorporates ammonia, on exposure to which, the **real image** is formed. Diazo film usually has a bluish cast to it and is difficult to scratch. Diazo film is a **nonreversing film** and is most often used for **microfiche**.

DIAZO TRANSPARENCY PROCESS. *See* **transparency**.

DIGITAL.
This usually refers to **data** which has been translated to, or captured in the form of, digits (and, most often, **binary** digits). Before data can be manipulated or otherwise worked with by a **computer**, it must be represented by (or expressed as) numerical **characters**. This process is often referred to as digitizing. Usually contrasted with **analog**.

DIGITAL COMPUTER. *See* **computer**.

DIGITAL COUNTER.
A device found in almost all **tape recorder**s of all types (i.e., audio **cassette, open-reel, videotape**, etc.). It usually counts the number of rotations of one of the **reel**s. The number can be seen through little windows somewhere on the surface plate and near the controls.
Most digital counters provide read-outs of either three or four digits. The latter offers greater accuracy when the operator is using the digital counter as an indexing device to find a specific portion of the tape. Digital counters that count only reel rotations provide only relative accuracy. More expensive tape recorders sometimes include a digital counter of greater accuracy, actually measuring the amount of tape passing a fixed point.

DIGITAL DISPLAY.
The means by which **data** can be displayed in discrete numerical values rather than by some analogous means. For example, a clock which presents the user with the numbers 12:15, etc., has a digital display as opposed to a clock whose hands point to the appropriate time. In the latter case, the clock utilizes an **analog** display.

DIGITAL READOUT. *See* **digital display**.

DIGITAL RECEIVER. *See* **tuner**.

DIGITAL RECORDING.
A means of recording **data** or **information** by representing them in a **digital** form. Although, historically, the **computer** was the principal reason for converting data to a digital representation, in the last several years the digital recording of **audio** and **video** information has become rather popular. Digital **tape recorder**s can now record music, etc., with far greater accuracy than the old **analog** method could, and without **distortion** or **noise** such as **tape hiss**. The **video disk** is a means for digitally encoding such material as television **program**s (in some systems, with **stereophonic** sound) with greater fidelity than can the **videotape recorder**.

DIGITAL TUNER. *See* **tuner**.

DIORAMA.
A three-dimensional display representative of a real-life scene. In museums of science or natural history they are usually constructed to life-size scale, but at the school level they are most often produced as miniatures.

DIRECT ACCESS. *See* **random access**.

DIRECT ACCESS STORAGE.
Storage of magnetically recorded **data** or **information** on a device which allows a specific item of data or information to be retrieved without the necessity of **read**ing all preceding data or information. **Disk** storage is direct access, while **magnetic tape** storage requires serial access wherein all items preceding the desired item must be read. Direct access storage device is often abbreviated DASD.

DIRECT BROADCAST SATELLITE.
A **communications satellite** which broadcasts a **signal** of sufficient strength that a small, inexpensive **antenna** and **receiver** suffice to utilize that signal. Such an antenna and receiver are suitable for use by private individuals.

DIRECT CONNECT MODEM. *See* **modem**.

DIRECT CURRENT (DC).
A flow of electricity that is constant in the direction of its flow (as opposed to **alternating current**). It is ordinarily produced by batteries—i.e., dry or wet cells—and by some components within electrical (or electronic) devices.

DIRECT DISC RECORDING. *See* **direct recording**.

DIRECT DRIVE.
The method of imparting motion from an electric motor to another component within a device, in contradistinction to **belt drive** or **rim drive**. The term *direct drive* is used in connection with such audiovisual devices as **turntable**s and **tape recorder**s. It tends to be a more expensive design than either the belt or rim drive, but it usually provides more accurate and reliable service. Direct drive tape recorders usually have three motors, rather than just one.

DIRECT MEMORY ACCESS.
A method of transferring **data** which employs a special piece of **hardware** and which sets up a high-speed path between the **computer memory** and **peripheral equipment** (e.g., a **terminal**).

DIRECT PRESSING. *See* **direct recording**.

DIRECT RECORDING.
The oldest method of **audio** recording (which, lately, has enjoyed some vogue) which makes no use at all of the **tape recorder**. Instead, a master **disc** is created (or cut) on a lathe whose **stylus** is activated electromechanically, but "directly," as the performance takes place. The advantages of direct recording are spontaneity (since **editing** later is impossible), no **tape hiss**, greater **dynamic range**, and relative freedom from **distortion**—all of the preceding assuming very careful work by both performers and engineers. Disadvantages are the necessity to re-record if an error is made (since, again, there can be no later editing),

DIRECT RECORDING (cont'd)

under- or over-**modulation** of the groove (unless the recording engineer follows the **program** carefully as he records) with resultant distortion, and very high prices since the master recording must be created in one "sitting." **Digital recording** offers many of the same advantages, but since it employs a digital **magnetic tape** master, the disadvantages are essentially nil.

DISC.

This spelling is herein used to differentiate the word from **disk**. The author has chosen the older form, *disc* (from the Latin *discus*) to represent a flat, round, plate on which are inscribed grooves in which are the **analog** representations of **data** or **information** (usually of an **audio** or **video** nature).

DISCRETE FOUR CHANNEL.

This term is properly used only when describing **quadraphonic** sound obtained from a **magnetic tape** that has been recorded with four separate **channels**. The **signals** from each channel are fed directly to four **amplifiers** (or two **stereophonic** amplifiers) and then to four **loudspeakers**. There is no need in this type of system for either a **decoder** or **demodulator**. The manufacturers of **CD-4** demodulators and recordings have also used "discrete four channel" to describe the CD-4 system. This system, however, is not truly discrete; the signals for the four channels have been merged (through the process of **modulation**) into two channels and are thus no longer truly discrete, or separate.

Discrete four channel tapes can be recorded and played back properly only on **tape recorders** having the proper **head** configuration, i.e., four track **record head**s and **playback head**s.

DISH.

An **antenna** whose shape is that of a parabolic dish. The dish antenna is ordinarily used to receive **signal**s emanating from a microwave source such as in telephone communication or from a **communications satellite**.

DISK.

In this work, disk is differentiated from **disc** in that disk is taken to mean a flat, round plate, covered with a special material by means of which a **magnetic recording** or other type of recording can be encoded on its surface in **digital** form.

DISK DRIVE.

The mechanism into which can be placed **data**-accepting **disks** or **disk pack**s and in which they rotate. The **head**(s) for **read** and **write** are on the arm(s) within the disk drive.

DISKETTE. *See* **floppy disk**.

DISK OPERATING SYSTEM (DOS).

An **operating system** which, in some small **computer**s, is stored on, and invoked from, **disk storage** to conserve **main storage** space.

DISK PACK.

A group of **data** disks, used as a **storage** medium which are usually enclosed in a container, and which can be mounted on a **disk drive**.

DISK STORAGE.
The **storage** of **data** on **magnetic disk**, rather than on **magnetic tape, core, punched card,** etc.

DISPERSION.
This specification pertains to **loudspeaker**s, indicating the distribution of sound over a stated angle throughout the listening room. Ordinarily, dispersion is given in degrees to indicate the arc of the horizontal plane through which sound emanating from the loudspeaker is evenly distributed. This type of even distribution is sometimes specified as effective dispersion.

Since the higher **frequency** sounds tend to be highly directional—i.e., they appear to emanate from a specific point in space—while lower frequencies are more generally dispersed (or omni-directional), many loudspeaker manufacturers use more **tweeter**s and **mid-range driver**s than **woofer**s in their loudspeaker systems in order to effect a greater dispersion.

Many critical listeners feel that too much dispersion (more than 90 degrees to 100 degrees) tends to cause a loss of **stereophonic** effect, while others feel that only with wide dispersion (160 degrees to 180 degrees) can real concert hall **ambiance** be achieved.

DISPERSION (LIGHT).
A variation of the refractive index of a substance with a **wavelength** (or color) of the light. It is because of its dispersion that a prism is able to form a spectrum. For most materials the **index of refraction (light)** increases as the wavelength decreases.

DISTORTION.
Any change, whether caused by the recording or by the reproducing equipment, from the original **audio** or **video** program. Distortion can be either electrical (particularly in audio equipment) or mechanical.

The most common types of audio distortion are **dropout, flutter, harmonic distortion, intermodulation distortion, phase reversal, rumble,** and **wow**.

With camera and **film** projection equipment, the types of distortion one may encounter in projection or viewing devices (other than in television and **videotape**) are **aberration,** projector **framing,** and **keystone effect**.

With television, videotape, etc., one may also find the following kinds of distortion: **D/C restoration,** snow **dropout,** and **roll-over**.

Systems that are hybrid and that combine both audio and visual devices (e.g., sound **motion picture projector**s) may suffer from distortion in either the sound or the film system, or in both.

There will always be some inherent distortion in any system that attempts to reproduce original programs, whether audio or visual. However, most distortion can be minimized to the extent that it is not perceivable or at least not obtrusive.

Most types of distortion are cited as a percentage of deviation from an ideal, or perfection, or from the total actual **signal**. For example, harmonic distortion may be given as .2 percent—that is, .2 percent of the total signal consists of harmonic distortion. The percentage of distortion that is intolerable or even noticeable varies depending on the type of distortion.

DISTRIBUTED PROCESSING.
A rather general term which is usually understood to mean the use of **smart terminal**s, **minicomputer**s or **microcomputer**s to do **data processing** on a decentralized basis away from (often with the help of, sometimes in lieu of) a **main frame**.

DOCUMENT FEED.

A means of feeding (or loading) single sheets of paper, **microfiche**, etc., into either a **copying machine** or **facsimile transmission** device. In most machines which have the document feed capability, this capability is automatic but of one of two types: A) single sheet feed, wherein an operator must hand-feed the device one sheet at a time, or B) stacked feed, in which the operator simply loads a specified number of sheets into a hopper (or other receptacle) and the device "feeds itself" unattended. This latter is often called "sheet feed."

DOCUMENTARY FILM.

A type of **film** that has a basis in fact and that concerns real-life situations and real people. However, all sorts of artistic techniques may be used to enhance the "realism" of the film or to convey the message or mood to the viewer.

DOCUMENTATION.

The **information** contained in such materials as **flowchart**s, narrative expositions, coding sheets, and operating instructions, which act as a reference manual for procedures, **program**s, etc. It is the documentation (which, in toto, may be thought of as a laboratory manual or diary) which allows one person to continue the work of another, or one person simply to understand another's programming or procedural efforts, or the workings of the programs, procedures, etc., themselves.

DODGING (PHOTOGRAPHY).

The technique of manually shading a portion of the **print** paper, while the **image** on the **negative** is being projected onto the print paper by the **enlarger**. The dodging process allows the dark room worker to lighten parts of the print which would otherwise be too dark.

DOLBY NOISE REDUCTION.

A system named for Ray Dolby, an American engineer who invented a specific technique for reducing unwanted **noise** in the **audio magnetic recording** process.

Three versions of the Dolby system are now available. "Dolby A," used by professional recordists, to date is not available in consumer form. "Dolby B" is less complex, but still quite effective; it is available in many audio **cassette** devices, **open-reel** type **tape recorder**s and **receiver**s (to "decode" **frequency modulation (FM)** programs that are "Dolby encoded"). "Dolby C" is more effective than Dolby B and is to be found on only the newest equipment.

The Dolby B version is designed to remove a great deal of the **tape hiss** inherent in audio magnetic recording. Many manufacturers of pre-recorded audio cassettes and **cartridge**s and open-reel tapes are now processing these tapes in Dolby B. This means that the user must have a device (e.g., an open-reel tape recorder) with the appropriate Dolby B circuitry in order to derive the best quality playback. However, separate Dolby B devices have recently been coming on the market and may be used with virtually any tape recorder. One can still play Dolby B encoded tapes on a device not so equipped if one reduces the **treble**, otherwise the sound will seem both hissy and overly bright.

Phonograph disc manufacturers are producing many discs made from master tapes that were Dolby processed. These discs, however, do not require that the user have special Dolby **decoder**s, and they have less obtrusive noise (like hiss) when played on regular phonographs.

DOLBY NOISE REDUCTION (cont'd)
Use of the Dolby system has measurably improved the sound quality of **magnetic tape**s, particularly those recorded at slower speeds, as in audio cassettes and cartridges.

DOLLY.
A platform, either on wheels or tracks, upon which a **motion picture film** camera is placed. The dolly allows the camera to be moved toward or away from a given scene. When the dolly is moved as the camera is running, the filmed result is termed a dolly shot.

DOT MATRIX PRINTER.
A **printer** or printing **terminal** which prints **character**s which are formed by a matrix of dots, either of ink (in an **impact printer**), or "hot spots" (in a **thermal printer**).

DOT PRINTER. *See* **dot matrix printer.**

DOUBLE EXPOSURE. *See* **multiple exposure.**

DOUBLE-PERFORATED FILM.
Roll film which has **sprocket** holes along both longitudinal edges.

DOWNTIME.
The interval of time during which a device is not operating because of **hardware** problems or **program** (or other **software**) **bug**s.

DRIFT.
An undesirable phenomenon of **frequency modulation** (FM) broadcast radio. It occurs as the components of an FM **tuner** warm up, and it can be discerned as a loss of optimum tuning—i.e., the station to which it is tuned appears weaker (less loud) and may be obscured by static.

Drift is usually not a problem in the best tuners. Less expensive tuners (or less well designed ones) use a special circuit (**automatic frequency control** circuit) to minimize or prevent drift.

DRIVER.
A single **piston speaker** in a system of **loudspeaker**s. There are basically three types of drivers: the **woofer**, the **tweeter**, and the **mid-range driver**.

DROPOUT.
A momentary loss of **signal** in a **magnetic recording**, usually caused by a brief loss of contact between the **head** and the **magnetic tape**. In **audio**, a dropout appears as a sound loss which lasts a fraction of a second. In **videotape**, dropouts can occur in either the sound or the **video** system. If in the sound system, the phenomenon is as described for audio recording. If in the video system, there may be very brief white streaks through the picture on the **cathode-ray tube**.

Dropouts are caused by either or both of two phenomena: dirty heads or poorly manufactured magnetic tape. In the latter case, the tape may not have been milled smooth enough, or the tape **coating** may be producing minute flakes, which interpose themselves between the head and the main portion of the coating. Most often, however, dropouts are caused by dirty heads.

DRUM.
A cylinder-shaped device, coated with a material which allows for **magnetic recording** of **data**. Drums were once a popular **storage** medium but were replaced by **magnetic tape** and **magnetic disk**s.

DRY CARREL. *See* **carrel**.

DRY MOUNTING OF PICTURES.
A process for mounting a picture, map, or facsimile document on a cardboard (or cloth) background; the process uses a special dry mount tissue and heat. (Pictures may also be mounted with rubber cement or liquid adhesives.) In mounting pictures, a piece of dry mount tissue is attached to the back of the picture in a few spots with a **tacking iron** (a small, wedge-shaped iron), and then the tissue is trimmed away from the picture with scissors. Next, the picture and the tissue are tacked to the mounting board in a few spots. A hand iron or dry mount press may then be used to firmly press all areas of the picture to the mounting board. If desired, the picture can be coated with a plastic spray or lacquer to protect it.

DRY TONER. *See* **toner**.

DUAL TRACK HEAD. *See* **half track**.

DUBBING.
The process of copying any previously recorded material—e.g., dubbing a **phonograph disc**'s program onto a **magnetic tape**, or copying the sound from an **open-reel** tape onto an audio **cassette** tape, etc.

Dubbing is also used to refer to the motion picture **film** process of adding dialogue (in **synchronization** with the projected film) when the originally recorded dialogue is either inaudible or in a foreign language. Occasionally dubbing refers to the process of adding music to the spoken dialogue on a film.

Over-dubbing is a term used to describe the **sound-on-sound** process in **magnetic recording**.

DUMB TERMINAL.
A **terminal** which has only the capability of transmitting or receiving **data** as it is keyed (or otherwise) fed in, i.e., acting as a simple **input/output** device. A dumb terminal will not store a **program** since, in effect, it has no **memory**. Terminals with greater capability are called **smart terminal**s and **intelligent terminal**s, the former being not as "bright" as the latter.

DUMP.
To copy the contents of a **memory** (or other **storage**) onto another memory, or storage, or onto paper. This allows for the contents either to be preserved or examined.

DUPLEX (COMPUTER).
The ability of two **terminal**s to transmit **data**, via a communications line, simultaneously to each other. This is often called "full duplex." "Half duplex" is the ability of but one terminal to transmit at a time, but the situation in half duplex can be reversed (i.e., A may transmit while B receives, then B may transmit while A receives) as contrasted to **simplex**.

DUPLEX (COPYING MACHINES).
The ability of a **copying machine** to print on both sides of a sheet of paper.

DUST COVER.
Just as its name suggests, a dust cover is a lid (often of rigid, transparent plastic) to be placed on such devices as **turntable**s and **tape recorder**s, designed to keep out dust.

DYNAMIC MICROPHONE.
A **microphone** that does not require its own supply of electrical power and that is somewhat more rugged (and less expensive) than the **condenser microphone**. Because of this, it has been rather more popular than the condenser microphone. Compared directly, a good condenser microphone offers better **frequency response** and **dynamic range sensitivity** than a good dynamic microphone. The dynamic microphone, however, suffices in any but the most exacting recording conditions; it is fine for classroom work, etc. Two of the most important specifications for any microphone are **frequency response** and **polar pattern**.

DYNAMIC RANGE.
The ratio (usually given in **decibels**) between the quietest portion and the loudest portion of a specific **program**. For example, the dynamic extremes of an orchestral concert of classical music may be 50 decibels to 115 decibels. The dynamic range, then, would be 65 decibels, but the specifics (50 decibels to 115 decibels) should be included.

Dynamic range is sometimes given as a specification for a piece of **audio** equipment. It refers to the equipment's capability of reproducing a given range of **amplitude** intensities with negligible **distortion**.

DYNAMIC SPEAKER. *See* **piston speaker**.

EBCDIC.
Acronym for Extended Binary Coded Decimal Interchange Code. Like **ASCII**, EBCDIC determines the number and sequence of **bits** for each **character**. EBCDIC uses an eight bit **code**. The only eight bit code competitor to it is ASCII-8.

ERIC. *See* **Educational Resources Information Centers**.

ETV. *See* **educational television**.

EARPHONE. *See* **headphone**.

EARTH STATION.
The point on earth at which a **signal** is either received from or sent to a **communications satellite**. The earth station also may feed a signal to (or receive it from) other terrestrial locations.

EASEL (PHOTOGRAPHY).
A device, either of wood or metal, made to hold a piece of unexposed **print** paper while under an **enlarger**. The easel should hold the paper very flat during the **exposure** in order to ensure edge-to-edge sharpness of the **image** when the print is **develop**ed.

EDGE NOTCH.
A system of encoding **information** on a card by punching notches along the edges of the card. This system has most often been used in libraries to encode circulation particulars such as due date, borrower's name, call number, etc. Because of the manual method of finding desired cards, the edge notch system is cumbersome, time consuming and outdated.

EDGE PRINTING.
The printing of **information** along the edge of a **film**, outside the normal **image** area (i.e., the **frame**). Edge printing usually conveys the name of the manufacturer, **film speed**, frame numbers, etc.

EDIT (DATA PROCESSING).
In a **word processor**, or similar **hardware**, the edit capability is usually one of **software** which permits the system to perform certain functions with minimal instruction. Some of these functions are changing the spelling of specified words throughout the text, creating an index and setting page sizes.

In **data processing**, generally, an edit may consist of preparing selected **data** for further processing involving validation, deletion, etc.

EDITING, FILM. See **splice, film**.

EDITING (RECORDING).
The deletion, addition, or rearrangement of portions of a recorded **program**, whether **audio** or **video**. In **magnetic recording**, editing can be accomplished in two ways: the **magnetic tape** can be cut and spliced (as is done with motion picture **film**), or it can be **erase**d and re-recorded. For editing in the second fashion, it is helpful if the **tape recorder** has a **pause control**.

When cutting and splicing magnetic tape, do not use blades, scissors, etc., that are magnetized, since this can produce a clicking or popping sound at the splice point. It is probably best to use a commercially available tape splicer, which has non-magnetizable metal cutting parts.

EDUCATIONAL FILMS. See **film**.

EDUCATIONAL RESOURCES INFORMATION CENTERS (ERIC).
A nation-wide series of eighteen clearinghouses that can provide **microfiche** or **hard copy** of documents and articles pertaining to education, such as reports of innovative **program**s, conference proceedings, bibliographies, outstanding professional papers, curriculum-related materials, and significant efforts in educational research and development. ERIC at Stanford University concentrates on educational technology, while other clearinghouses in the system specialize in such phases of education as reading and communication skills, rural education and small schools, educational management, and exceptional children. The ERIC system can be searched either by **computer** or manually, but one should first refer to the *Thesaurus of ERIC Descriptors* and then go to *Research in Education* (RIE) or *Current Index to Journals in Education* (CIJE). RIE includes abstracts on all processed documents except journal articles. CIJE indexes relevant journal articles and includes brief annotations.

EDUCATIONAL TECHNOLOGY. See **instructional technology**.

EDUCATIONAL TELEVISION.
A general term used in reference to non-commercial television operations. **Public television, school television,** and **instructional television** are classified as educational television.

EDUCATIONAL TELEVISION, CLOSED CIRCUIT.
A transmission system that distributes television **program**s, live or tape, both **audio** and **video**, to a limited **network** connected by cable. The network may consist of one school, a whole school district, or several districts. The telecast cannot be received by other TV sets outside the selected network. The **signal** does not have to meet U.S. **Federal Communications Commission** commercial broadcast specifications.

EFFICIENCY.
In **audio** this refers to the amount of electrical power (measured in watts) necessary for a **loudspeaker** system to produce audible sounds. Less efficient systems, like the **acoustic suspension** type, require more power than, say, bass reflex systems (see **baffle**). Efficiency is not a measure of the quality of reproduction, but only of the amount of power the loudspeaker requires. Therefore, a **high-efficiency** loudspeaker is not necessarily better than a **low-efficiency** loudspeaker, and vice versa.

8mm FILM. *See* **film**.

EIGHT TRACK.
This is applied as a descriptive term to audio **cartridge** tapes. It indicates that these **magnetic tape**s contain, usually, four **stereophonic** programs (two **track**s per program).

ELECTRET CONDENSER MICROPHONE. *See* **condenser microphone**.

ELECTROFAX COPYING.
A type of **electrostatic copying** in which the transfer step is eliminated because the **image** is placed directly on the copy paper. Since this paper contains a photosensitive coating, the technique is often called coated paper copying. The paper is, obviously, more expensive than plain bond copying paper and may not be easily written on with a ball point pen or regular pencil. Otherwise, the image on it, although not necessarily of archival quality, is essentially as good as that produced by other electrostatic copying techniques.

ELECTRON BEAM RECORDING.
A method of producing **computer-output microform** in which an electron beam is used to **write output data** or **information** directly onto the microform.

ELECTRONIC MAIL.
The transmission of messages (excluding **graphic**s) within a system (usually using a **word processor** or **computer**) from the originator to the intended recipient without the necessity of first committing the message to paper. Usually the sender enters the message on a **keyboard** and the recipient is able to view it on a **cathode-ray tube** or obtain it as a **printout**.

ELECTRONIC OFFICE.
An office, group of offices, or institution, the personnel of which have access to **terminal**s which are connected to a central system (either a **word processor** or **computer**) which enables them to obviate the use of paper to a great extent. Some of the capabilities of

ELECTRONIC OFFICE (cont'd)
the electronic office include the use of **electronic mail**; the preparation, including **edit**ing, of texts without paper; and the creation of **data files** (kept usually on **floppy disk** in the case of small word processing based systems, or any larger **storage** device in the instance of larger systems).

ELECTRONIC TUNER.
In a radio or television **receiver**, a device for selecting a station or **channel**. It has no moving parts and selects the desired station or channel directly, that is, without having to tune intervening stations or channels. For example, if a television set is tuned to channel 13, channel 7 can be tuned without having to pass either through channels 2 through 6 or 12 through 8. Electronic tuners are more reliable, efficient, quieter, and longer lived than are mechanical ones.

ELECTROSTATIC COPYING.
A **copying machine** system which uses a **medium** covered by a photoconductive surface which can accept (and hold) a positive electrical charge. This medium (often in the shape of a large cylinder or "drum") is exposed to the **image** on the original document. Light which is reflected off areas of the drum where the original casts no image, negates (or "erases") the positive charge (leaving a charge only where the image appears). **Toner** (which may be in either powder or liquid form, depending on the manufacturer's design) is flowed onto the surface of the drum and, since the toner is negatively charged, clings to the areas carrying the positive charge. A sheet of paper (usually bond paper except in **electrofax copying**) passes against the drum surface and between it and a current carrying wire which causes the toner to adhere to the paper. Prior to delivery to the user, the paper is passed through a heater ("fuser") which fuses the toner to the paper.

ELECTROSTATIC HEADPHONE.
A **headphone** that operates on the same principle as an **electrostatic speaker**, as opposed to the dynamic, or **piston speaker**. Electrostatic headphones are usually more expensive than the piston type, but their **frequency response** is better.

ELECTROSTATIC MICROPHONE. *See* **condenser microphone.**

ELECTROSTATIC PRINTER. *See* **electrostatic copying**.

ELECTROSTATIC SPEAKER.
A type of **loudspeaker** that differs from the common **piston speaker** in its method of producing sound. The electrostatic speaker has a **diaphragm** that produces the sound, rather than a **cone** or piston. The diaphragm is usually a flat metalized plastic (often mylar), which is suspended between two acoustically transparent metal plates and which is activated by a static electric field instead of a magnetic field. This field activates the entire diaphragm, rather than just a portion of it, which avoids the creation of unwanted sounds. However, electrostatic speakers have some difficulty in reproducing sounds below about five hundred **hertz** and are frequently used in combination with **piston speaker** type **woofer**s. In general, the piston speaker is still the best type for all-around institutional use.

ELLIPTICAL STYLUS.
A **stylus**, used in **phonograph**s, whose cross-section is an ellipse and whose tip is rounded. It is designed for better reproduction of **stereophonic discs** than can be obtained by using a **conical stylus**, but it is usually more expensive. The conical stylus is adequate for

ELLIPTICAL STYLUS (cont'd)

reproduction of 78 rpm or **monaural** discs. The best stylus for stereophonic and **quadraphonic** reproduction of discs is the **Shibata stylus**, or one of its variations.

EMULATE.

A situation in which a given **computer**, outfitted with special **hardware** (and sometimes **software**) can perform work whose **program**s were originally written for a different type of computer.

EMULSION.

A gelatinous layer of photosensitive material on **film** in which the **latent image** is formed.

ENCLOSURE.

A cabinet in which a **loudspeaker** (or loudspeaker system) is housed. The enclosure's material and dimensions are critical and vary with the specific characteristics of the loudspeaker or system.

ENLARGEMENT (PHOTOGRAPHY).

A **print** which is made from a relatively small **negative** and which is considerably larger than that negative. The enlargement process requires the use of an **enlarger**. When an enlargement is made from a **microform**, it is frequently referred to as a **blowback**.

ENLARGEMENT RATIO.

The difference in linear size between a **negative** and an **enlargement** made from it. Supposing the enlargement to be 15 **diameter**s greater, the enlargement ratio can be expressed either as 15x (the "x" indicating the multiplication factor) or 1:15.

The inverse process, that is the making smaller of an **image**, is called reduction and employs the concept of a "reduction ratio."

ENLARGER.

A device used to increase the size of the **image** from a relatively small **film negative** when making a **print**. It projects the image onto photosensitive paper.

The enlarger is often used as a means of enhancing the compositional quality of the negative. This is done by allowing undesirable portions of the image on the negative to flow beyond the paper on which the picture is to be printed, so that these portions do not appear in the picture. This process is one form of **cropping**.

ENLARGER/PRINTER. *See* **reader/printer**.

EQUALIZATION.

In **audio**, this refers to the adjustment of **frequency response**. In commercially made **disc** recordings, the manufacturer boosts the high **frequency** portion of the **frequency range** to mask the **tape hiss**, while the low frequency portion may be compressed, in **amplitude**, in order to save groove space. Equipment used to play back these recordings will de-emphasize the high frequency boost and emphasize the low frequency cuts proportionately. The amount of equalization in the recording and playback equipment has been standardized, since about 1955, by the **Recording Industry Association of America**. Most recording manufacturers in the United States follow these standards, as do those equipment manufacturers who retail their equipment in the United States.

EQUALIZER. *See* **frequency balance control.**

ERASABLE PROGRAMMABLE READ ONLY MEMORY (EPROM).
Similar to the **programmable read only memory (PROM)** except that by means of strong ultraviolet light (or, recently, by electrical means) the EPROM can be **erase**d and **reprogram**med. Contrasted with **read only memory** and **random access memory.**

ERASE.
In **magnetic recording** this is a process of removing a **signal** from the **magnetic tape** by passing the tape next to the **erase head**, while the erase head generates a strong random (rather than specific) magnetic field. Most **tape recorder**s automatically activate the erase head when functioning in the ordinary recording **mode** to eliminate (erase) previously recorded signals.

ERASE HEAD.
The first in a series of **head**s in a **magnetic recorder**. It **erase**s previously recorded **signal**s on the **magnetic tape**, thereby preparing the tape to receive new **program** material.

ERROR RATE.
The ratio between the total number of **character**s (or other **data** units) sent and those among that total which were incorrect due to any faults in transmission.

EUROPEAN STYLE PRINT. *See* **matte.**

EXCITER LAMP.
A lamp used to scan the **optical sound track** of **film**s as they run through **motion picture projectors**. It activates the **photoelectric cell.**

EXECUTIVE PROGRAM.
A system of relatively complex routines, resident in **main storage**, which controls the order in which other **program**s are loaded and run.

EXPOSURE.
In photography, exposure may mean any one of the following: a **negative** which, having been exposed to the subject of the photograph, now carries a **latent image**; the amount of time a negative has been exposed to the subject, by means of the **shutter** (more properly "exposure time"); or, lastly, the number of **frame**s on a **film** which have yet to be exposed.

EXPOSURE INDEX.
The number which some standardizing agency assigns to a photosensitive material (like **film**) and which indicates its relative sensitivity to light in order that the **aperture**, **shutter** speed, etc., are properly set. (In the United States, the **American National Standards Institute** sets the exposure index for photosensitive materials.)

EXPOSURE METER.
An instrument, usually sufficiently small as to be hand held, which measures the light either falling on, or reflected by, a photographic subject. If it measures the light falling on a subject, the meter is said to measure "incident light"; if it measures the light reflected by the subject, "reflected light" is what is measured. Most modern cameras, particularly those termed automatic, incorporate some type of light meter which is coupled to either the **shutter** or **iris** (or both) and which assists in the proper setting of them.

78 / EXTERNAL STORAGE

EXTERNAL STORAGE. *See* **auxiliary storage.**

FCC. *See* **Federal Communications Commission.**

FE. *See* **field engineer.**

FET. *See* **field effect transistor.**

FM. *See* **frequency modulation.**

FORTRAN.
　A **high-level language** designed for mathematical, engineering and scientific use. The acronym *FORTRAN* is formed from "formula translation." FORTRAN is probably the most widely used language for **computer program**s of a technical nature.

FPS. *See* **frames-per-second.**

f/NUMBER. *See* **lens speed.**

f/RATING. *See* **lens speed.**

f/STOP. *See* **lens speed.**

FACSIMILE.
　An exact copy (both in **image** detail and image size) of an original document. A fine facsimile, in other words, should be a counterfeit (in the best sense).

FACSIMILE TRANSMISSION.
　The transmission of an **image** of a document, from one point to another, by electronic means (usually telephone). The image is sent by a **source terminal** called a transmitter and is received by a terminal termed a **sink terminal.** The received image, when reconstructed onto paper by the sink terminal, is the **facsimile.**

FALSE DROP.
　Sometimes called a false retrieval, a false drop is erroneously retrieved **data** or **information** from a **computer.** False drops usually occur when the material desired is improperly described or specified, or when, as in searching a **data bank,** it was not properly indexed prior to being **input.**

FAST FILM. *See* **film speed.**

FAST FORWARD.
　Mode in a **tape recorder** in which the **magnetic tape** is quickly passed from the **supply reel** to the **take-up reel.** During fast forward the tape should not be in contact with the **head**s since, at this fast speed, the tape can cause excessive head wear. The opposite of fast forward is **rewind.**

FAST LENS. *See* **lens speed.**

FAX. *See* **facsimile.**

FEDERAL COMMUNICATIONS COMMISSION (FCC).
An independent executive agency of the United States government which is authorized to regulate radio, television, telephone, and telegraph operations within the United States. It allocates the specific **frequency** and **channel** for different types of communication activities and is empowered to grant, revoke, renew, and modify broadcasting licenses.

FEEDBACK. *See* **acoustic feedback** and **programmed instruction**.

FEEDBACK (CONTROL).
Any device, group of devices, or system which samples and measures conditions in order to modify them to predetermined levels. The thermostat is among the simplest, and most common, feedback devices. Other examples are **servomechanism**s and photoelectric controls which automatically turn on electric lighting when predetermined levels of darkness occur.

FEED REEL. *See* **supply reel**.

FELTBOARD. *See* **clothboard**.

FEMALE PLUG. *See* **jack**.

FERRICHROME TAPE.
A **magnetic recording** tape with a new **coating** formulation. Ferrichrome (frequently FeCr) ostensibly enjoys the advantages of both **ferric oxide tape** and **chromium dioxide tape**. Supposedly, it records the high **frequency range** well, and is less abrasive than chromium dioxide tape.
However, very few machines are now manufactured which utilize ferrichrome tape and it will probably be abandoned. It is **tape type** III.

FERRIC OXIDE TAPE.
A **magnetic recording** tape whose **coating** is made of ferric oxide (Fe_2O_3, or simply FeO). FeO tape has been in use for about thirty years, and is the most common of **magnetic tape** formulations. It is **tape type** I.
In recent years, other formulations have been placed on the market, particularly **metal tape**.
For general use, FeO tape is essentially adequate, and it is certainly the least expensive of the three.
Ferric oxide, by the way, is very close, in formula, to ordinary rust, so it is a good practice to clean the **head**s, **capstan**, and other parts of the **tape recorder** regularly to remove the abrasive detritus of the FeO tapes that have passed through the mechanism.

FIBER OPTICS.
The transmission of light (which light often carries **data**) through specially designed long, thin fibers of glass or plastic. Such fibers transmit data far more quickly and at much higher density than can metal cables carrying electrical current.

FIELD.
In **data processing**, the specific portion of a **data** record allocated to a unit of **information**. As an example, a circulation record may have separate fields for due date, call number, borrower's name, etc.

FIELD EFFECT TRANSISTOR (FET).
This is a transistor made of silicon and silicon oxide and frequently used in **tuner**s and other types of equipment because of good linearity and high **impedance** of **input**.

FIELD ENGINEER.
The engineer or technician of a specific manufacturer who is capable of making on-the-spot repairs, etc.

FILE LAYOUT.
The organizational format (and specifics of the types and format of contents to be found there) of a **computer** file. In other words, a description of the arrangement, and kind, of **data** to be found within a computerized file.

FILE MAINTENANCE. *See* **maintenance**.

FILM.
Films for educational and commercial uses are produced in several sizes: 8mm, 16mm, 35mm, and 70mm. The "mm" stands for millimeters, i.e., width of the film in millimeters. If they are sound films, they will be provided with a sound track on one side of the film, which will be either an **optical sound track** or a magnetic sound **stripe**. The largest sizes (35mm and 70mm) are most widely used for professional motion pictures, while the smaller sizes (8mm and 16mm) are used in schools and for home movie making by amateurs. The 16mm size has long been used in schools, but in recent years the 8mm film has found wide application in education, especially the 8mm **film cartridge** (in either standard or Super 8mm forms). Super 8mm size film and the projection equipment to go with it were brought out by the Eastman Kodak Company in 1965. Super 8mm produced a better **screen image** (because of the 50 percent larger **frame** size) than the standard 8mm film. Also, a magnetic sound stripe was added on one side of the film. Using a projector with sound capabilities allows teachers and students to record their own sound tracks and, in effect, to become film-makers.

Films that are used in the field of education have been variously entitled educational films, teaching films, or instructional films. Most of these have been designed and produced specifically for the classroom, but in some instances commercial or entertainment films can also be used in courses such as social studies, English, or science. However, these commercial films have to be specially printed for use in 16mm **motion picture projectors**.

Films can be used in a number of ways to achieve educational objectives, such as communicating **information**, changing or strengthening attitudes, developing skills, evoking curiosity, raising problems, or setting moods. Some educational films are exclusively instructional and are designed to explain a process or develop a skill. Representative of this type is the single-concept film, which is uniquely appropriate for the 8mm continuous film loop or film cartridge format.

Most school district, county, or single school **media center**s have catalogs or lists that describe films available for classroom use. Large numbers of 16mm films can be rented directly through college, university, and commercial rental libraries.

FILM ANIMATION. *See* **animation, film**.

FILM CARTRIDGE.
A standard 8mm or Super 8mm **film** loop of short length or duration, which is sealed in a plastic container in such a way that the beginning and end of the film are joined. An

FILM CARTRIDGE (cont'd)
example is the Super 8mm Technicolor cartridge which carries a three-minute film in an endless loop. Some 8mm cartridges have sound recorded on the film loop by means of a magnetic **stripe**, while others have an **optical sound track**. There are also 8mm cartridges that do not have sound. For viewing, the film cartridge is simply inserted into a slot in specific 8mm **motion picture projectors** designed for cartridges, and the machine is turned on. A great advantage of film cartridges is that they require no threading or **rewind**ing. Some types of **rear projection** 8mm film cartridge equipment are prefocused and turn off automatically when the film comes to an end. Film cartridge is also called film cassette.

FILM CASSETTE. *See* **film cartridge.**

FILM CHAIN.
A group of components such as **motion picture projectors, slide projector,** TV camera **pickup tube,** and controls that compose a system to provide **video** information from motion picture **film** or from **slide**s.

FILM EDITOR. *See* **splice, film.**

FILM FRAME. *See* **frame.**

FILM GATE. *See* **gate.**

FILM GRAIN.
The individual particles of a photosensitive silver compound found in a **film**. Because these particles are of a somewhat randomly varying size, distribution, and **sensitivity,** film grain causes a fuzzy (or grainy) appearance in both **negative**s and **print**s. A general rule is the greater the **enlargement ratio,** the grainier the enlargement.

FILM LEADER. *See* **leader.**

FILM LOOPS. *See* **film cartridge** and **motion picture projectors.**

FILM MASTER.
The **microform negative** produced in the camera and often referred to as the master or film master.

FILM NEGATIVE. *See* **negative.**

FILMOGRAPHIC TECHNIQUE.
A process of using still **print**s, photographs, paintings, or drawings in a **film** to create the illusion of motion.

FILM PATH. *See* **motion picture projectors.**

FILM PLANE.
An imaginary line drawn at right angles to the length of a camera lens or **film** projector lens, and through which the film moves.

FILM PLANE SHUTTER. *See* **focal plane shutter.**

82 / FILM PROJECTOR

FILM PROJECTOR. *See* **motion picture projectors.**

FILM SPEED (MOTION). *See* **frames-per-second.**

FILM SPEED (SENSITIVITY).
A rating of the sensitivity of a **film**'s **emulsion** to light—commonly called its film speed (sensitivity)—given as an ASA number. The higher the ASA number, the "faster" (more sensitive to light) the film emulsion.
The ASA numbers are arithmetical in their progression, thus the doubling of an ASA number indicates a film emulsion of twice the sensitivity. For example, ASA 400 is four times as sensitive as ASA 100, but only twice as sensitive as ASA 200.

FILMSTRIP.
A series of half **frame** or full frame pictures on a strip of 35mm **film**, sometimes called a **slide** film. They may be in color or black and white. Early filmstrips were primarily full frame, but in recent years the half frame format has become standard. Some filmstrip projectors may be multiple-use machines—that is, they can also be used for showing slides. Multiple-use machines usually have attachments for slide viewing and the projector need not be operated manually, but can be run by remote control. However, many filmstrip projectors require that the film be advanced manually.
Filmstrip viewers, small devices into which the user inserts the filmstrip, project the **images** on the filmstrip by means of **rear screen projection**. Since such viewers are usually small, they are ideal for individual use, as in a "wet" **carrel**. Some filmstrip viewers are so small that they are **hand viewers**. These may or may not incorporate their own light sources.
Filmstrips are frequently designated as either "sound" or silent. In fact, there is no difference between the two, since filmstrips do not carry sound information in the form of either an **optical sound track** or a magnetic sound **stripe**. Sound filmstrips simply have an accompanying audio **cassette** or **phonograph disc**, which usually includes either narration and/or musical background. They may or may not incorporate an audible (or inaudible) **signal** which, when used with the proper equipment, automatically advances the filmstrip to the next frame.

FILMSTRIP PROJECTOR. *See* **filmstrip.**

FILMSTRIP VIEWER. *See* **filmstrip.**

FILTER (PHOTOGRAPHY).
A material (usually of glass with a gelatin or plastic layer, or of a dyed plastic like cellophane) which is translucent but only transmits light of certain **wavelengths** (or colors). For example, a red filter will allow only the red portion of the spectrum to pass. Two other types of common photographic filters are the **ultraviolet** (or UV) filter and the polarizing filter. The former filters out most ultraviolet light (the principal cause of clouds appearing "washed out" or too light) while the latter reduces glare from highly reflective surfaces like water and glass. Care must be used with virtually all filters (except the ultraviolet filter) to correct for the **filter factor**.

FILTER FACTOR.
The use of almost any photographic **filter** decreases the amount of light passing through the lens to the **film**. In order to compensate for this light loss, most filters are assigned a filter factor, by means of which factor the **shutter** speed or **aperture** may be set.

FILTER SWITCH.

This is found on most **audio** integrated **amplifiers** and **preamplifiers**. The circuit(s) that such switches activate are designed to reduce the **amplitude levels** in the high **frequency** and low frequency sections of the **audible frequency range**.

Most **tape hiss** and **phonograph disc** "scratchiness" occurs above about eight thousand or ten thousand **hertz**. The high filter, therefore, causes diminution of frequencies above that level.

The **rumble** or **hum** that can occur in audio devices or **software** tends to be below one hundred hertz, so the low filter usually diminishes frequencies below that point.

The higher filter is occasionally called the scratch filter and the low filter may be called either the **bass** or the **rumble** filter. In any event, they should be used judiciously; when these circuits have been activated, it sometimes sounds as if a large portion of the audible frequency range has been deleted—thereby interfering with one's listening pleasure.

FIXED FOCUS.

Labels a camera or projector which does not allow for some focus change in order to compensate for differences in distance between the subject and the camera or the **screen** and the projector. Usually a fixed focus device is less expensive (and desirable) than one with variable focus.

FIXED LENGTH.

A **data** record or **field** which must contain a specific number of **characters**. Contrasted with **variable length**.

FLAG.

A piece of **data** or **information** (usually but a **bit**) carried with a longer item of data or information and conveying something informative regarding that to which it is attached. For example, a flag may indicate that the piece of data to which it is attached is to be treated as both a person's driver's license number and social security number. Flag is often used interchangeably with "tag."

FLANNELBOARD. *See* **clothboard**.

FLAT.

A common term in **audio** vocabularies, used to describe a **frequency response** whose **amplitude** does not vary at all. If such an idealized frequency response were to be graphed, it would produce a perfectly straight horizontal line—ergo, flat. Flat may also refer to setting the **bass** and **treble** controls on an audio device in a neutral position, thus ostensibly neither adding to nor subtracting from the frequency response of the **program**.

FLAT-BED CAMERA. *See* **planetary camera**.

FLATS. *See* **glass flats**.

FLETCHER-MUNSON CURVES.

Named for the two people who devised them, these curves describe the Fletcher-Munson Effect, which calculates the loss of audibility to the human ear of the low **frequency range** as **amplitude** is decreased. This means that as the volume **level** is decreased, we seem to hear less of the **bass**. This apparent bass loss is due to characteristics of the human ear, since **audio** measuring devices do not detect this ostensible loss of low frequencies.

FLETCHER-MUNSON CURVES (cont'd)

Most audio manufacturers build into their **receivers, amplifiers,** etc., a "**loudness compensation**" circuit, which, when activated by the appropriate switch, will boost the amplitude of frequencies in the bass range proportionately to the overall volume setting in order to provide the human ear with what seems a normal **frequency response**.

FLIP-FLOP. *See* **roll-over**.

FLOCKBOARD. *See* **clothboard**.

FLOP-OVER. *See* **roll-over**.

FLOPPY DISK.

A flexible **magnetic disk** (usually of mylar) coated with a surface for **magnetic recording** of **data**. Floppy disks vary in size, usually 5¼-inches (diskettes) or 8-inches and may be recorded on one or both sides. The floppy disk is usually used as a **storage medium** for the **minicomputer** and the **microcomputer**.

FLOWCHART.

The logical representation, in the form of a diagram, of the possible situations to be encountered in a given process. For example, a flowchart may be made of the circulation process in a specific library. In it, situations (and the potential outcomes or responses) such as the borrower not possessing a library card, the book not being a normal circulating copy, etc., can be depicted. Flowcharts usually use standard symbols (and templates with these symbols are available), and the flowcharts themselves are useful prior to attempting to convert a human process or procedure to one handled by a **computer**. This is because the flowchart allows the person in **systems analysis** to determine whether such a conversion is feasible or desirable.

FLUORESCENT DISPLAY.

A set of **fluorescent light** bulbs used as **signal strength meter**s, **digital counter**s, clocks, etc., in **audio** or **video** equipment.

FLUORESCENT LIGHT.

Artificial light created in a special tube (or lamp) by means of an electrical current which ionizes mercury vapor trapped in the tube and causes it to radiate. The radiation activates a fluorescent material coated on the inside of the tube and causes it to emit even greater light. Fluorescent light is considerably closer to **natural light** than is **incandescent light**, but tends, because of its diffuseness, to cast fewer shadows than either natural or incandescent light.

FLUTTER.

A specification appropriate to **tape recorder**s and **phonograph turntable**s. Flutter, manifested as a rapid wavering (or fluttering) of pitch, is usually a form of mechanical, rather than electrical, **distortion**. Flutter is due to speed changes, which are caused by any of a number of mechanical faults: the puck (rubber rim) in a **rim drive** system may have a few flat or worn spots; the belt in a **belt drive** system may be slipping; a motor may be varying in speed; etc.

When flutter is cited as a specification, it is given as a percentage of deviation from perfect speed; the lower the figure, the better. Sometimes flutter and **wow** are combined as a specification. If flutter is given by itself, it should not exceed .05 percent for turntables or

FLUTTER (cont'd)
tape recorders. A combined flutter and wow specification for a turntable should not exceed .15 percent; for a tape recorder (**open-reel**), .1 percent at 7½ **inches-per-second tape speed**; and for an audio **cassette** device, .2 percent.

FOCAL LENGTH.
The distance between the optical center of a lens and the **film plane**, when the lens is focused on infinity. The focal length determines the size of the **image**. The greater the focal length, the larger the apparent size of a distant object. A telephoto lens is one with a long focal length, while a wide-angle lens has a comparatively short focal length. The focal length is almost always specified in millimeters.

FOCAL PLANE SHUTTER.
A type of **shutter** used most often in the **single lens reflex camera**. The focal plane shutter consists of one or more curtains of metal or cloth which lie just in front of (and parallel to) the **film plane**. A slit in the curtain(s) allows light to pass from the lens to the **film**. The amount of time the shutter allows light to strike the film (i.e., the **shutter speed** setting) is usually determined by how quickly the slit moves through the film plane. Some cameras, however, may have more than one slit, with differing widths, the specific width (and the speed with which it moves) are determined by the camera operator.

The focal plane shutter can offer faster speed settings than the **between-the-lens shutter**, and it tends to be more accurate in its speed. However, the focal plane shutter is often considerably noisier and, because of its design, it may cause **distortion** of objects moving relative to the camera.

FOIL TAPE SENSING.
A system incorporated into some **open-reel** type **tape recorder**s which allows the device to sense a piece of metal-foil tape applied to a certain point on the **magnetic tape**; when the metal-foil tape is sensed, the device goes into **automatic rewind** or **automatic reverse mode**, depending on the machine and how it has been **program**med to react.

FOLDED DIPOLE.
An **antenna** frequently used for non-fringe (i.e., close) **frequency modulation** (FM) broadcast reception. Its length should be approximately thirty inches, and it can be made of flat television antenna **lead-in** wire (three hundred ohm). If an outdoor antenna designed for FM reception is not used, a "rabbit ears" type of television antenna will usually provide better reception than a simple folded dipole.

FOOT CANDLE.
A measure (or unit) of illumination. One foot candle is equal to one **lumen** per square foot of area illuminated. The term has as its origin the intensity of light at a distance of one foot from a "standard" (i.e., one inch diameter) candle.

FOOTPRINT.
The area of the earth's surface on which the **signal** from a **communications satellite** may be received. With an ideally placed satellite, its footprint can cover about one-third of the globe's surface.

FOUR CHANNEL. *See* **quadraphonic**.

FOUR TRACK.

A **magnetic tape** recorded by a **quarter track** design **open-reel** type **tape recorder** or a **stereophonic** audio **cassette** tape recorder with four **track**s of recorded **information**. Four track is not to be confused with four **channel** which is appropriate to **quadraphonic** sound.

FOUR TRACK HEAD.

A **playback head** or **record head** devised to playback or record the **quadraphonic** program on an **audio** type **magnetic tape**; that is, four **track**s simultaneously.

FRAME.

A single picture in the series printed on a length of **film**. It also refers to the rectangular shape that bounds the **view finder** field-of-view of a camera and hence, the picture that will eventually be shown on the **screen**.

Frame is also used to define the small units of subject matter that are the basic format of **programmed instruction**.

FRAME (TELEVISION).

One complete picture consisting of two sections of interlaced **scanning lines** within the **raster**.

FRAMES-PER-SECOND (fps).

This indicates the running speed of **film** in a camera or projector. The normal 16mm speed for silent films is 16 fps, and for sound films it is 24 fps. 8mm film speed is either 18 fps or 24 fps. Super 8mm film speeds are the same as regular 8mm, except that the faster speed is usually used for sound films.

FRAMING (PROJECTOR).

All **motion picture projectors** are fitted with a control that adjusts the position of the picture in relation to the **pawl-sprocket** mechanism. This is necessary to ensure that the **frame** is correctly centered in the **aperture** so that the picture fills the **screen** properly—that is, so that it is cut off neither at the top nor at the bottom.

FREEZE-FRAME. *See* **stop-motion**.

FREQUENCY.

This refers to the oscillations of a wave or of a **signal** expressed in **hertz**. Hertz were formerly described as "cycles-per-second," which meant that a given signal emanates as a wave that vibrates so many times per second. For instance, in **audio**, the note A (to which American orchestras tune) has a frequency of 440 hertz (or 440 cycles-per-second).

Described musically, frequency doubles, or halves, with the octave. Therefore, the A above the one cited above has a frequency of 880 hertz, while the A below it has a frequency of 220 hertz. The audio frequencies that a human ear can hear are usually termed the **audible frequency range**.

Frequency is also used to describe any flow that has regular periodicity. An example would be **alternating current**, which changes its direction of flow sixty times a second, in the United States, and is said to have a frequency of 60 hertz.

FREQUENCY BALANCE CONTROL.

An **audio** device designed to complement a regular audio system. It allows a finer control (often octave by octave) of the **audible frequency range** that ordinary **tone controls** can. With this device, the audio system can, in effect, be "tuned" to the listening

FREQUENCY BALANCE CONTROL (cont'd)
environment. It is particularly useful in listening environments that are acoustically peculiar and may also be used to improve a poor recording in the **dubbing** process.

Frequency balance controls are sometimes called "equalizers," "frequency equalizers," "frequency spectrum equalizers or balancers," or "graphic equalizers."

FREQUENCY EQUALIZER. *See* **frequency balance control**.

FREQUENCY MODULATION (FM).
A form of radio broadcasting that uses a **carrier signal** modulated in **frequency**. In the United States, the FM **band** extends from 88,000,000 **hertz** to 108,000,000 hertz.

FM broadcasting is often referred to as **high fidelity** broadcasting in that its sound quality is generally better than that offered by **amplitude modulation** (AM) broadcasting.

For various reasons, FM offers a better **frequency range** (about 30 hertz to 12,000 hertz) and is less subject to **interference**, either from electrical devices or from natural phenomena (like lightning) than AM. However, FM waves travel in straight lines, unlike AM waves, which follow the curvature of the earth. This means that, as the distance between the FM transmitter and **receiver** increases, there is some loss of broadcast quality.

Stereophonic type **program**s broadcast in FM are usually termed **multiplex**. Such broadcasts are compatible with receivers and **tuner**s that are **monaural**; this means, for instance, that a multiplex FM program can be received, and listened to, in an automobile equipped with a monaural FM radio.

FREQUENCY RANGE.
A **frequency** group with lower and upper limits, is usually given in **hertz**. For instance, the **audible frequency range** is usually given as approximately forty hertz to eighteen thousand hertz. It differs from a frequency **band** in that a frequency band is ordinarily used to describe a frequency range specific to broadcasting, like the "short wave" band.

FREQUENCY RESPONSE.
A basic **audio** specification, it has two components—one is a **frequency range**, and the other is the deviation, in **amplitude**, along the frequency range. For example, the frequency response of a **tuner** may be given as thirty **hertz** to thirteen thousand hertz (this is the frequency range), plus or minus 1.2 **decibel**s (this part is the amplitude deviation).

The frequency response reveals whether or not the tuner is capable of reproducing the stated frequency range, and it indicates that, in the example above, no given **frequency** will vary in volume (amplitude) more than 1.2 decibels louder or softer. This means that there is a potential total of 2.4 decibels variance. When two or more frequency response specifications are being compared, several things must be remembered. One is the **audible frequency range** (approximately forty hertz to eighteen thousand hertz), and the other is: the lower the amplitude deviation, the better. For instance, comparing the following two frequency responses—sixteen hertz to twenty-seven thousand hertz plus or minus 3.5 decibels, and twenty-four hertz to twenty thousand hertz plus or minus .9 decibels—the better of the two is the second, since it covers a good deal of the audible frequency range with considerably less amplitude variation.

A minimum frequency response for tuners, **amplifier**s and **receiver**s should approximate ten hertz to twenty-five thousand hertz plus or minus .5 decibels. For **tape recorder**s (**open-reel**), the minimum (at tape speed of 7½ inches-per-second) should be thirty hertz to sixteen thousand hertz plus or minus 2 decibels. For audio **cassette** machines, a frequency response of fifty hertz to thirteen thousand hertz plus or minus 3 decibels would be desirable. A **phono cartridge** should have a frequency response from forty hertz to twenty thousand hertz plus or minus 3 decibels.

FREQUENCY SPECTRUM BALANCE. *See* **frequency balance control**.

FRONT-SURFACED MIRROR.
A mirror whose reflective surface is on the front, rather than the rear (as is most common). Front-surfaced mirrors are used in photography (and other optical systems) because they eliminate the possibility of double **image** formation by light rays which are not parallel to the axis of the mirror.

Most front-surfaced mirrors are not protected by some harder, additional material. For this reason they must be cleaned very carefully (usually with a soft brush, e.g., camel's hair) to prevent scratching.

FULL DUPLEX. *See* **duplex (computer)**.

FULL FRAME. *See* **slide**.

FULL-SIZED COPY. *See* **hard copy**.

FULL TRACK.
A **magnetic recording** of a **monaural** type **program** on **open-reel** style **magnetic tape**. The **track** is recorded for almost the full width of the tape, and in but one direction, as opposed to **half track** recording. This is a system essentially no longer in use.

FUNCTION SELECTOR. *See* **input selector**.

GIGO. *See* **garbage-in-garbage out**.

GAIN.
This refers to an increase or decrease in electrical **signal** strength (or **amplitude**). Providing gain is the function of an **amplifier**, and the **volume control** is sometimes referred to as the gain control.

GANGED CONTROL.
Either a single control (like "volume" on an **amplifier**) which acts on both **stereophonic channel**s, or two or more controls on one spindle or shaft. In the second case (for instance, **treble** controls, again on an amplifier), the control knobs may be concentric, one fitting into the other and each acting on a separate channel.

GAP.
The tiny space between the pole pieces of the electromagnet in a **head**. It is at the gap that the **erase**, record, or playback functions actually take place. Because the gap is so small (it is measured in microns), it is easily blocked by material flaking off the tape **coating**. Therefore, it is necessary to clean the heads carefully and regularly.

GARBAGE-IN-GARBAGE-OUT.
An old **data processing** or **computer** hand's aphorism which loosely translates: If the **data** or **information** put into the system resembles a sow's ear, don't expect technology to transmute it into a silk purse.

GATE.
The component in a camera or projector that holds each **frame** of the **film** flat and momentarily still behind the lens.

GAUSS.
A unit of magnetic induction named for K. F. Gauss. It is appropriate to **magnetic recording**, in that the **playback head** and **record head** in most such devices require periodic **degauss**ing.

GENERATION.
As with living things, generation (usually preceded by an ordinal number) is an indication of how many steps removed from an ancestor (i.e., the original) a descendant (i.e., a copy) may be. For instance, a copy of an original document made by **electrostatic copying** is a first generation copy (as are all copies made directly from the original). Copies made from the first generation copy are second generation, etc. The same is true for any **medium** which lends itself to such copying (e.g., **magnetic tape, film,** etc.). It is important to remember that the greater the distance the copy is from the original (i.e., the higher the generation number), the greater the denigration of the original exhibited by the copy. In film (or, for that matter, copies of documents made on **copying machine**s) there is a loss of distinctness and, ultimately, there are unclear **image**s. In **audio** and **video** recording there is increased **distortion** or **dropout**s.

GEOSTATIONARY SATELLITE. *See* **communications satellite**.

GHOST (TELEVISION).
A shadowy or weak **image** in the received picture, offset either to the left or right of the primary image; it is the result of transmission conditions that create secondary **signal**s, which are received earlier or later than the main or primary signal. Sometimes the result of **multipath**.

GLASS FLATS.
Pieces of glass which have been carefully polished to remove almost all surface irregularities (therefore, flats) and between which **film** is placed in order to view or copy it in an **enlarger, reader** or **reader/printer**.

GLITCH.
A technical problem, usually small and relatively quickly remedied, in an electronic device. Different from a **bug** in that a glitch is almost always **hardware** related. A glitch is nowhere near as catastrophic as a **crash**.

GLITCH (TELEVISION).
A form of low **frequency interference**, appearing as a narrow horizontal bar moving vertically through the picture.

GLOSSY.
A photographic **print** having a highly glazed (or shiny) appearance (especially when contrasted with a **matte** surface print). To obtain a glossy, a special print paper (one with a glossy **emulsion**) is used. After processing, the print is finished by pressing its emulsioned surface against a very smooth metal plate (the plate's surface usually is coated with enamel or may be chromium plated; or the plate itself may be highly polished copper, stainless steel or glass) called a ferrotype plate (not to be confused with a ferrotype print, which was a print made on a ferrotype plate covered with its own photosensitive emulsion, similar to a tintype).

A glossy print has a hard, bright surface compared to the soft, dull one of a matte print. The glossy is said to be of high **reflectance**.

GRAIN. *See* **film grain**.

GRAPHIC.
A diagrammatic or pictorial representation as opposed or contrasted to narration or text. The ability to create graphics by a **computer** or other **data processing hardware**, such as a **word processor**, is a special one and not all hardware are appropriately equipped. There has to be attendant specialized **software** plus devices like a **plotter**, in order for the system to be able to create graphics. The graphics capability is a very useful one for the pictorial representation of statistics, maps, etc.

GRAPHIC EQUALIZER. *See* **frequency balance control**.

GRID. *See* **microfiche grid**.

HD. *See* **harmonic distortion**.

HD TELEVISION. *See* **high definition television**.

Hz. *See* **hertz**.

HALATION.
The word *halation* is etymologically derived from halo and is almost self-explanatory. In photography, halation is a kind of halo (or diffused or blurred effect) around the main **image** of a fairly bright source of light. It is caused by light reflecting from the **film** backing or **base**. It most often appears in photographs made at night.

Most modern films now have anti-halation bases which greatly reduce halation.

HALF DUPLEX. *See* **duplex (computer)**.

HALF FRAME. *See* **slide** and **filmstrip**.

HALFTONE.
The reproduction of a photograph (or other picture) in which the gradation of tone (i.e., from white to varying gray to virtual black) is obtained by photographing the work to be printed through a screen which divides the **image** into small dots which appear farther apart or closer, depending on how dark portions of the original are. From this screened photograph, an engraved plate is made which is then used to **print** the picture. Virtually all photographs in newspapers (and most black-and-white photographs in books) are reproduced via the halftone method. To determine whether this is so, simply examine such a picture with a magnifying glass.

HALF TRACK.
A **magnetic recording** of either a **monaural** or **stereophonic** type **program** on **open-reel** type **magnetic tape**. Each of the two recorded **track**s occupies slightly less than one-half the width of the tape. On a monaural tape the tracks may run in opposite directions, each containing one **channel** of **information**. When one side of a monaural half track tape has been recorded or played, the tape must be turned over to record, or play, the track of the remaining side. In half track stereophonic recording or playback, the two tracks run in the same direction; one contains left channel information; the other, right channel.

Half track tapes are frequently used in professional recording studios or for broadcast. Tapes so recorded employ a half track head; therefore, they cannot be used on **full track** or **quarter track tape recorder**s because they are not compatible with these **head** configurations.

HALO (TELEVISION).

This usually refers to a dark area surrounding an unusually bright object caused by an overload of the camera **pickup tube**. Reflection of studio lights from a piece of jewelry, for example, might cause this effect. With certain camera pickup tube operating adjustments, a halo shows as a white area surrounding a dark object.

HAND READER. *See* **hand viewer**.

HAND VIEWER.

A small and portable device (capable of fitting the hand) by means of which **microform**s may be read. Most hand viewers are specific to a given microform (e.g., a **microfiche** hand viewer) and do not carry their own light sources, but must be used in conjunction with an external light source (e.g., the sun, a lamp, etc.).

HARD COPY.

A copy of a document which is, as opposed to a **microform**, legible to the naked eye. It is used somewhat interchangeably with "paper copy."

A more accurately descriptive term would have been full-sized copy.

Hard copy should also be distinguished from a legible display as, say, on a **cathode-ray tube**. A "hard copy" may be made of that display, but would properly be termed a **printout**.

HARDWARE.

Electrical, electronic, or mechanical devices or equipment (as distinct from the media suitable for them, these media being termed **software**). For instance, a **tape recorder** is a piece of hardware, while the **magnetic tape** for it is **software**. Likewise, a **computer** is hardware and a **program** is software.

HARD WIRE.

A communication line which connects **data processing** equipment and which is **dedicated** to that system. This in contrast to the **dial-up** approach. A hard wire is not ordinarily part of a larger communications system, but exists only to serve the devices to which it is connected.

HARMONIC DISTORTION (HD).

A **distortion** specification most often applied to **loudspeakers**, **tuners**, **amplifiers**, **receivers**, **tape recorders**, and **preamplifiers**. Harmonic distortion means that the device for which the specification is given takes an **input signal** of given **frequency** (or frequencies) and adds to it other frequencies, which are exact multiples of the original frequency. The specification is expressed as a percentage of the total **output**. Thus, an amplifier may be cited as having a harmonic distortion of .3 percent.

One aspect of harmonic distortion is that it tends to decrease as **power output** increases until the full power output of the device is approached, at which point harmonic distortion rises dramatically.

Of late a more popular means of expressing harmonic distortion is as total harmonic distortion (THD). It is a somewhat more complex, but more accurate, way of expressing this type of distortion. It, too, is expressed as a percentage. Total harmonic distortion for a device other than a loudspeaker should not be greater than .15 percent. Obviously, the lower the figure the better.

HEAD.
The most elemental component of a **magnetic recording** device. Basically, the head is a very small electromagnet (or electromagnets in one housing, as in a **quarter track** or **half track** type **stereophonic** head). There are three types of heads: the **erase head**, the **record head**, and the **playback head**. The portion of the head that actually **erase**s, records, or plays back is the tiny **gap** between the magnet's pole pieces.

Some **tape recorder**s (the usual audio **cassette** type is an example) contain two heads: one is for erase and the other is a combined record/playback head.

Many **open-reel** tape recorders have separate erase, record, and playback heads, arranged in that order. In this way, the operator can **monitor** the actual recording process and make necessary adjustments as the recording is being made. Machines with but two heads ordinarily do not provide this capability of **tape monitoring** the recording process.

Some tape recorders even have four or more heads to allow playback, or erase/record/playback in the opposite direction, eliminating the need to turn the tape over to record or play back on the other side of the tape.

Heads are particularly subject to wear when cheaper brands of tape, which may be very abrasive, are used. Even good tape can shed its **coating** particles onto the heads. Therefore, it is a good idea to clean the heads with a cotton swab dipped in commercially available head cleaner or isopropyl alcohol at least once a month.

Further, the record and playback heads build up, through use, a magnetic charge that can permanently damage both the heads and the tapes passing over them. Therefore, these heads should be **degauss**ed periodically.

Lastly, the alignment of the heads at right angles to the longitudinal axis of the tape is critical. After long use, the alignment of the heads should be checked — but only by a really knowledgeable technician or repairman.

HEAD CRASH.
A mechanical or electrical failure of a **computer**'s **disk drive** which results in physical contact between a **head** and the rotating disk.

A head **crash** causes damage to both the head and the magnetic **coating** of the disk, usually necessitating replacement of both and, usually, causing the loss of most or all **data** stored on the disk.

HEADER.
The eye-legible portion on a **microform**. Most commonly found in **microfiche**, it is almost always at the top or front ("head") and usually contains bibliographic material like an author's name, document title, and pagination. The header's primary use is for filing and identifying the microform.

HEAD GAP. *See* **gap**.

HEADPHONE.
This consists of small **loudspeaker**s, usually in pairs, mounted to a headband so that each loudspeaker fits over an ear. Headphones are designed for individual listening to **audio** sources. Some headphones use **piston speaker**s; the very best ones use electrostatic **transducer**s (*see also* **electrostatic headphone**).

Older headphones tended to close the listener off from all sounds other than those emanating from the headphones themselves. A newer design allows the user to hear external sounds, also. These newer headphones are known by a variety of descriptive terms, including hear-through, high velocity, and dynamic velocity. Good headphones will have a **frequency response** from about 30 **hertz** to sixteen thousand hertz plus or minus two **decibel**s.

HEADSET. *See* **headphone**.

HEAD SHELL.
That part of the **tone arm** of a **phonograph's turntable** which houses the **phono cartridge**. In better turntables, the head shell is detachable.

HEAT SINK.
A device (usually of finned metal) which draws unwanted heat away from a heat sensitive device (like a transistor) and dissipates that heat into the air.

HELICAL HEAD.
A record/playback type of **head** used in many **videotape** and **video cassette** recording devices. The head is mounted in a drum that allows the head(s) to rotate in such a way that the **track** recorded on the tape is a section of a helix. The helical head concerns only the recording or playing back of the picture, not the **audio** portion of the **program**. Helical heads are quite delicate, and their alignment is critical; they should be treated with great care.

HERTZ (Hz).
The unit of **frequency** equal to one cycle-per-second. Named for H. R. Hertz, the hertz unit replaced the older, more descriptive term "cycles-per-second" in 1967.
An **audio** example of the use of the term is the representation of the musical note A above middle C; this note equals 440 hertz.

HEURISTIC.
A problem solving technique which uses a trial-and-error method. The heuristic approach is usually contrasted with the one which employs **algorithm**s.

HI-FILTER SWITCH. *See* **filter switch**.

HIGH CONTRAST. *See* **contrast**.

HIGH DEFINITION TELEVISION.
A television system in which 1,100 or more **scanning line**s are utilized to produce the picture, as contrasted with the 525 lines standard in television systems in the United States.

HIGH-EFFICIENCY.
A type of **loudspeaker** system that requires comparatively less electrical power to create sound, as distinguished from **low-efficiency** systems, like the **acoustic suspension** system.
Ordinarily, a high-efficiency system can produce room-filling sound with ten or fifteen watts of electrical power, while a low-efficiency one may require up to thirty watts.
Efficiency, it should be noted, is not a measure of quality—it is merely descriptive of power requirements and consumption.

HIGH FIDELITY.
This term should not be confused with **stereophonic**. High fidelity has come to mean **audio** system **hardware** and **software** that produce sounds very close to the original sounds, whether broadcast or recorded.

HIGH FILTER. *See* **filter switch**.

HIGH LEVEL. *See* **level**.

HIGH-LEVEL LANGUAGE.

A **language** which is used to write a **computer program** and which employs statements in "natural language" (e.g., English) which correspond to several **machine language** instructions. The functions performed in response to the commands are usually consistent with the natural language meaning of the command. High-level languages also tend to be more application or problem oriented while a **low-level language** tends to be more machine or procedure oriented. An example of a high-level language is **FORTRAN**.

HISS. *See* **tape hiss**.

HISS SWITCH. *See* **filter switch**.

HIT.

In **data processing** parlance, a hit is a successful matching of two or more items. An example of a hit for a librarian is an appropriate bibliographic citation obtained through a **computer** literature search of a **data bank**. Hit is the opposite of **false drop**.

HOLLERITH CARD.

A standard **punched card** (sometimes called an IBM card) whose dimensions (7 3/8 x 3 1/4-inches) and **code** (in the form of punched holes) were designed (and later patented) by Dr. Herman Hollerith for the 1890 census. He went on to found the company which eventually evolved into IBM.

The Hollerith card consists of eighty vertical columns and twelve horizontal rows (the top three of which are called zones). All common English language alphabetical (i.e., A through Z) and Arabic numeral **character**s may be encoded as an **alphameric** set on a Hollerith card. The only limitation in the amount of data encoded on such a card is that it cannot exceed eighty characters (the number of vertical columns).

HOLLERITH CODE. *See* **Hollerith card**.

HOLOGRAM.

A photographic **emulsion** on a **film** or plate on which laser light has been used to record **information** from which a **virtual image** of the recorded scene can be reconstructed. The film or plate is simultaneously illuminated by light coming directly from a laser, and by light from the same laser reflected by the holographic subject onto the film or plate. After **develop**ing, the film or plate appears to have simply been accidently exposed to light: there is no apparent image. However, when the emulsion is exposed to laser light, a three dimensional image of the original scene may be viewed, as though one were looking through the film or plate at the original scene.

HOOK 'n LOOP BOARD. *See* **clothboard**.

HORN.

A type of **loudspeaker** "coupled" to the air by a large trombone-bell-like device. The horn type of loudspeaker is ordinarily of **high-efficiency**, capable of producing very good **treble** tones and often having very good **dispersion**.

HOST COMPUTER.

The controlling (or principal) **computer** in a system in which two or more computers are interconnected.

HUE. *See* **chroma**.

HUM.
An unwanted low **frequency** tone, usually continuous, and almost always at sixty **hertz**. It is due to problems caused by improperly grounded electrical and electronic **audio** equipment.
Hum can sometimes be eliminated (or minimized) by reversing the **plug**s of the **line cord**s of the equipment, that is, reversing these plugs relative to their **jack**s, assuming that these plugs are not of the grounded, three prong type.

HYSTERESIS MOTOR.
A type of motor used in better quality **turntable**s and **tape recorder**s. Its design allows operation at constant speed despite minor changes in **alternating current** power voltage.

IBM CARD. *See* **Hollerith card**.

IC. *See* **integrated circuit**.

IHF POWER.
The **power output** specification for an **amplifier** devised by the Institute of High Fidelity. This specification is poor because it indicates only the amount of power (in watts) an amplifier may deliver in "peaks" of short duration. A much better measure of power output is **continuous power**. IHF Power is similar to (and sometimes called) peak power or music power.

IMC (INSTRUCTIONAL MATERIALS CENTER). *See* **Media Center**.

IMD. *See* **intermodulation distortion**.

IMS. *See* **information management system**.

I/O. *See* **input/output**.

IPS. *See* **inches-per-second**.

ISO. *See* **International Organization for Standardization**.

ITFS. *See* **instructional television fixed stations**.

ITV. *See* **instructional television**.

ICONOSCOPE.
A television camera **pickup tube** in which a high velocity electron beam scans a **photoemissive** mosaic that has an electrical storage capability.

IDIOMORPHIC PROGRAMMING. *See* **computer assisted instruction**.

IDLER WHEEL.
A device in a **rim drive** type **phonograph**'s **turntable** which transmits the motion from the motor to the **turntable platter**.

96 / IDLER WHEEL

IDLER WHEEL (cont'd)

The idler wheel, sometimes called puck, is a disc with a rubber rim (hence puck) mounted on an axle. This rim touches both the shaft of the motor and the inside rim of the turntable platter.

The problem with this type of system lies in two properties of the idler wheel. The first is that the idler wheel's tension is maintained by a spring which, because of the rim drive design, has a tendency, with time, to stretch and lose its tension. The second is the rubber rim of the idler wheel. It can develop flat spots, become worn, or become hard and glazed. Either one of these conditions—spring tension loss or changes in the shape or properties of the rubber rim—can cause the idler wheel to rotate the platter at eccentric or improper speeds.

Belt drive or **direct drive** systems are definitely preferred to the rim drive system.

IMAGE.

One of three kinds of optical representation of a subject: **latent image**, **real image**, and **virtual image**. A latent image is one which is captured and encoded in a photosensitive material but which needs to be chemically, mechanically, or magnetically treated in order to be viewed by the naked eye. A real image is one formed by the convergence of light rays that have passed through an image-forming device (i.e., a lens); it can be projected onto the surface of a **film** or **screen**. A virtual image is one in which the light rays appear to diverge; it cannot be projected onto a film or screen.

IMAGE ORTHICON CAMERA TUBE (TELEVISION). *See* **orthicon image, camera tube** and **pickup tube**.

IMAGE REJECTION.

A specification of a radio **tuner** or **receiver**'s ability to reject a **signal** that appears at an incorrect point on the radio dial.

Image rejection is given in **decibel**s, and the higher the number the better. A minimum for **frequency modulation** (FM) radio is sixty-five decibels.

IMAGE ROTATION.

The ability of a **reader** or **reader/printer** to rotate (in a two dimensional plane) a projected **image** through a specified arc (usually at least 180 degrees) in order for the image to appear right side up.

IMAGING.

In **audio** terms, this is a function of the **dispersion** of **loudspeaker**s. Imaging is actually a subjective term attempting to describe the apparent locations (in three dimensions) of musical instruments, voices, etc.

IMPACT PRINTER.

A **printer**, often of a higher speed than the **thermal printer** type, which produces the symbol **image**s through a mechanical process (as distinct from chemical, thermal or photoelectric processes). The impact printer is one in which each **character** of the type is behind a printing medium (i.e., an inked ribbon, or some such) and the paper on which the printing occurs is in front. The characters are struck from behind by hammers (here is the "impact") which force them to press against the inked medium and, thus, imprint the paper. One form of impact printer is the **line printer**.

IMPEDANCE.

The opposition a circuit offers to the flow of an **alternating current**; it is measured in ohms. Impedance may vary with the **frequency** of the alternating current.

Loudspeakers offer impedance to the electrical current flow of the **amplifier**s to which they are connected. Usually, a loudspeaker's impedance will be four ohms, eight ohms, or sixteen ohms. Modern amplifiers can safely provide current to these impedances, but below four ohms, an amplifier could either burn out some of its **output** transistors or blow out its fuses.

INCANDESCENT LIGHT.

Artificial light, different from **fluorescent light** in that an electrical current is forced through a metal wire element (the filament) and the resistance of the filament to the current is sufficiently great to cause it to glow (or incandesce). The filament is housed within a glass bulb from which most air has been evacuated (causing a partial vacuum) or into which an inert gas such as nitrogen has been placed.

Incandescent light tends to have more of the red (and less of the blue) light of the spectrum than does **natural light** or fluorescent. Therefore, color **film** used in incandescent light and without some **filter** correction will tend to exhibit more brown through yellow hues than otherwise. Incandescent light also casts more shadows than fluorescent light because the bulbs are smaller.

INCHES-PER-SECOND (ips).

The measure of **tape speed** past a stationary point.

The most common speeds are: 15 ips (for professional recording and playback), 7½ ips (many home and institutional **tape recorder**s offer this for high quality music recording and playback), 3¾ ips (for lower quality music recording and playback; this is also the speed at which audio **cartridge** devices operate), and 1⅞ ips (for recording and playback of spoken material; also the speed at which audio **cassette** devices operate).

In general, the faster the tape speed, the better the potential **frequency response** of the recording, and the less **tape hiss** encountered.

INDEX COUNTER. *See* **digital counter**.

INDEXED FILM.

Microform film, usually **microfilm**, which uses lines, bars, or other such devices which enable the user to find a specific section of the film while scanning it rapidly on a **reader** or **reader/printer**.

INDEX OF REFRACTION (LIGHT).

A ratio of the sine of the angle of incidence to the sine of the angle of refraction, for a ray of light passing through the surface separating two media. It is also equal to the ratio of the velocity of light in the first **medium** to that in the second.

INDUCED MAGNET CARTRIDGE. *See* **magnetic cartridge** (3).

INDUSTRIAL CAMERA (TELEVISION). *See* **vidicon camera tube**.

INFINITE BAFFLE. *See* **baffle**.

INFORMATION.

Data which, having been collected, arranged, and displayed, convey meaning and may even impart intelligence or knowledge.

INFORMATION AREA (FILM).

Exclusive of **edge printing**, that portion of a **film** (usually used in reference to **microforms**) which contains the **information** to be conveyed. Ordinarily this is taken to mean the **frame** or **header**.

INFORMATION MANAGEMENT SYSTEM.

A system, usually involving **data processing**, in which **information** has been organized and cataloged and which provides **storage**, **retrieval** and **access**. Not to be confused with a **management information system**.

INFORMATION RETRIEVAL.

Any **computer** system which enables the user to locate and recover specific **information** from among a mass of **data**. The system from which the information is obtained is termed **information storage**.

INFORMATION STORAGE.

The **computer's storage** of **information** in the form of **data**. Locating and recalling this information is accomplished by **information retrieval**.

INFRARED.

That light whose rays are located slightly beyond the red end of the visible spectrum and extend to the shortest microwaves. Infrared is emitted by hot bodies and can be photographed using special **film, filters**, etc.

INPUT.

This is used as either a noun or a verb. In both instances it describes the concept (related either as a device or as a process) of transference of a **signal** from one device to another. The **output** signal from a **phono cartridge** is an input signal to another device; the phono cartridge itself is an input device, whose signal is fed to a **preamplifier** or **amplifier**. Further, one may input a signal into a **tape recorder** by singing into its **microphone**.

Most **audio** or visual systems have many potential input sources. An audio amplifier may be designed to accept inputs from a **phonograph**, a **tuner**, a tape recorder, a microphone, etc.

A **videotape** recorder may accept a camera input or a microphone input, or it may use a television set as an input in order to record broadcast programs, etc.

In order to operate properly, there must be electrical **compatibility** between the output of one device and the input of another. For example, the phono cartridge should have an output voltage that is matched to the input voltage of the amplifier or preamplifier with which it will be used.

Input is also the act (or the material engendered by the act, e.g., **data** or instructions) of entering material from a **terminal** or other **peripheral equipment** into **memory**.

INPUT/OUTPUT (I/O).

A **terminal** capable of entering **input** and receiving **output**.

INPUT SELECTOR.

The control on an **amplifier** or **preamplifier** which allows one to select from a variety of **program** sources (e.g., **phonograph, tuner, tape recorder**).

INSTANT REPLAY.

The ability of a **videotape recorder** to play back, immediately after recording, a section of the recording. Instant replay is most commonly associated with television broadcasts of sporting events in which events of particular interest are replayed, often in **slow motion**.

INSTITUTE OF HIGH FIDELITY. *See* **IHF power**.

INSTRUCTION, PROGRAMMED. *See* **programmed instruction**.

INSTRUCTIONAL FILMS. *See* **film**.

INSTRUCTIONAL MATERIALS CENTER. *See* **Media Center**.

INSTRUCTIONAL TECHNOLOGY.

A general term that includes the use of educational **hardware** and **software** for achieving specific objectives in the classroom; but it goes beyond this definition to encompass a systematic way of designing, implementing, and evaluating the total process of learning and teaching. In effect, it is one factor in a **systems approach** to education that attempts to determine which resource or combination of resources (people, places, media) is appropriate for teaching what type of subject matter to what type of learner under what conditions (time, place, size of group, etc.) to achieve what purpose.

INSTRUCTIONAL TELEVISION (ITV).

Television **program**ming specifically designed for the student, whether in the classroom or not. Its overall context pertains to the subject matter to be found in the courses of study of formal education.

INSTRUCTIONAL TELEVISION FIXED STATIONS (ITFS).

A number of **channel**s in the 2,500,000,000 **hertz band** which provide television opportunities for educational purposes; these channels have been set aside by the U.S. **Federal Communications Commission** for use by schools in areas approximating large school districts. The ITFS is a low-powered, limited range system which is less costly than **very high frequency** and **ultra high frequency** installations for transmitting **program**s. Its **signal**s are private because home television sets cannot receive them, and a school must have a special receiving system **(translator)** to convert the ITFS signal for distribution on its closed-circuit **channel**s.

INTEGRATED AMPLIFIER. *See* **amplifier**.

INTEGRATED CIRCUIT (IC).

A miniaturized electrical (or electronic) circuit assembly with certain of its components reduced to microscopic size. Such a device may be smaller than a thumbnail yet house the equivalent of hundreds of transistors, etc. The term *integrated* is used because the device's components are inseparable and formed on (or within) a continuous material. Integrated circuits have allowed the development of varying degrees of miniaturization in virtually all electronic **audio** and visual devices.

INTELLIGENT TERMINAL.

A **terminal** which is not only an **input/output** device but which also contains its own **memory** (less than a **smart terminal** and considerably more than a **dumb terminal**) and is capable of maintaining a modest **program** in order to provide local processing.

INTERACTIVE SYSTEM.

A **computer** system which operates in a conversational or interrogative fashion. That is, it asks and answers questions in order to elicit sufficient background to be able to provide the user with desired **data** or **information**. By this interactive refinement of the user's needs, the system is able to satisfy the user with greater accuracy. An interactive system (sometimes, previously, called an iterative system) requires a more complex **program** than one which simply responds without posing options or refining questions.

INTERFACE.

A shared boundary, often in the form of a person or device. The interface is capable of reacting to both terrains which it serves. A librarian, for example, may be the interface between a library user and a **data bank**. Another interface may be a **black box** which links two **computer** systems.

INTERFERENCE.

An undesirable influence on a **signal** caused by other signals. For instance, the motors of an airplane may cause interference with the reception of television **program**s (either the **video** or **audio** portion, or both).

INTERLOCK. *See* **record interlock**.

INTERMITTENT MECHANISM. *See* **pawl-sprocket**.

INTERMODULATION DISTORTION (IMD).

A specification, ordinarily applied to such **audio** equipment as an **amplifier, preamplifier, tape recorder,** and **tuner**. Intermodulation **distortion** is the measure of the mixing of one **frequency** with another to produce other undesirable frequencies. For example, a **signal** of 6,000 **hertz** and one of 200 hertz may produce signals of 6,200 hertz and 5,800 hertz.

Intermodulation distortion is expressed as a percentage of the total **output**, and the lower the percentage, the better.

One of the curious aspects of intermodulation distortion (an aspect that it shares with **harmonic distortion**), is that as the **power output** is increased, the intermodulation distortion tends to decrease, until, as the full potential power output is approached, the intermodulation distortion rises rapidly.

Good intermodulation distortion for a device other than a tape recorder would be less than 0.5 percent; for a tape recorder, 2.0 percent or less.

INTERNAL STORAGE.

Computer storage of **data** in the **central processing unit**'s **memory** rather than in **auxiliary storage**.

INTERNATIONAL ORGANIZATION FOR STANDARDIZATION (ISO).

An organization of national standards bodies representing each of the eighty-eight member nations. ISO publishes International Standards, and maintains a reference library of members' national standards, and is located at 1 rue de Varembe; Case Postale 56; CH-1121 Geneva 20; Switzerland.

INTRINSIC PROGRAMMING. *See* **programmed instruction**.

INVERTED FILE.
Any file, whether **computer**, catalog card, etc., which is not arranged by main bibliographic entry (i.e., personal or corporate author, title, etc.) but rather some other characteristic like a subject.

IRIS.
An adjustable **diaphragm** found in still and motion picture cameras and **enlargers**. It is usually made up of a series of overlapping metal blades, much like the blades in a **between-the-lens shutter**. Like this type of **shutter**, too, the iris is usually found within the structure of the lens. The iris, which forms the **aperture**, along with the shutter, controls the amount of light passing through the lens.

ITERATIVE SYSTEM. *See* **interactive system**.

JACK.
A socket or receptacle into which a **plug** is inserted. It is sometimes mistakenly called a female plug.

JACKETED FILM.
Strips of **film** housed in an acetate carrier which is composed of one or more sleeves. Such film is usually a **microform**, often **microfilm** which, when jacketed, resembles a **microfiche**.

JITTER (TELEVISION).
The tendency toward lack of **synchronization** of the picture. It may refer to individual lines in the picture or to the entire field of view.

JUSTIFICATION.
The marginal alignment of printing on a page. Right/left justification runs a line of print to a set margin on both edges of the page; right justification aligns print to the right hand margin only, left justification does the inverse. If only one margin is justified, it is usually the left one. Justification is often a feature in a **computer program, word processor**, or **terminal**s other than **dumb terminal**s.

kHz. *See* **kilohertz**.

KWIC. *See* **keyword**.

KWOC. *See* **keyword**.

KALVAR. *See* **vesicular film**.

KEYBOARD.
An electronic/mechanical **data processing** device (in appearance rather like a typewriter) which generates the **code** for a **character** when a key is depressed. A keyboard is almost always an adjunct to a **terminal** and is intended to **input** the **data**.

KEYPAD.
A **keyboard** small enough to fit in one's hand.

KEYPUNCH.
A piece of **peripheral equipment** (including a **keyboard**) which punches the holes in a **punched card**.

KEYSTONE EFFECT.
An optical distortion that occurs with the use of any type of projector when the lens is not parallel to the **screen**. It results in a broadening of the picture at the top and a narrowing at the bottom. This can be corrected by tilting the screen toward the projector (or vice versa); inexpensive specialized screens can be obtained that have been designed for this purpose.

KEYWORD.
A meaningful or informative word used for purposes of indexing and **retrieval** and usually taken from the title or abstract of a document. Keywords are usually selected by a **computer program** to produce an index also compiled by the computer. There are essentially two types of keyword indexes: key-word-in-context (KWIC) and key-word-out-of-context (KWOC). The KWIC index (sometimes called a permuted index) presents the keywords in columnar fashion on the page, while the titles in which the keywords are found are "wrapped" around the keywords. KWOC indexes list the keywords either at the left margin or in the center of the page, and present the appropriate titles in natural order beneath the proper keywords.

KEYWORD-IN-CONTEXT. *See* **keyword**.

KEYWORD-OUT-OF-CONTEXT. *See* **keyword**.

KILOHERTZ (kHz).
A combination of the multiplication factor one thousand, plus the term **hertz**. Therefore, five kilohertz equals five thousand hertz.

KINESCOPE.
A term frequently used to mean television picture tubes in general. However, this name has been copyrighted. Kinescope is also used to mean a motion picture **film** obtained by directly photographing the **image**s on a television picture tube.

LCD. *See* **liquid crystal display**.

LED. *See* **light-emitting-diode**.

LED READOUT. *See* **light-emitting-diode**.

LSI. *See* **integrated circuit**.

LAMINATE.
A piece of material (e.g., a **print**) which is overlaid with another material which is usually protective and often transparent.

LAMP.*See* **projection lamp**.

LANGUAGE, COMPUTER.
A set of rules governing the organization of a set of **characters** or symbols comprehensible to a **computer** and designed to represent an operation or set of operations. A computer may use a **high-level language, low-level language,** or **machine language.**

LANTERN SLIDE. *See* **slide.**

LATENT IMAGE.
The **image** on an undeveloped **exposure** (i.e., a **negative** or **print** which is not yet a **processed film** or print) which is not visible to the eye. Upon processing, the latent image becomes a **real image** or a **virtual image.**

LATITUDE (PHOTOGRAPHY).
The tolerance, or extent of **exposure** time, a given **film** or **print** paper allows while still producing an acceptable **image.** A film of higher **film speed** or a black-and-white film tends to have greater latitude than do slower film speed films or color films, respectively.

LAVALIER MICROPHONE.
A small **microphone** that can be suspended from the user's neck by a wire or cord or attached to his clothing with a clip. It is frequently used in broadcasting since it is unobtrusive and allows the wearer relative freedom of movement.

LEADER.
A section of plain tape or **film** that is added to the beginning of a **magnetic tape** or motion picture film. It is nonmagnetic, nonlight sensitive, and carries no **audio** or visual **information.** Its function is to provide a material tough enough to withstand the repeated rigors of the initial threading process.

LEAD-IN.
An insulated wire that interconnects an **antenna** and a **tuner, receiver,** or television set. Most lead-in wire is one of two types: that flat, two-conductor kind used with many black-and-white television sets, which has an **impedance** of three hundred ohms; and the round, **coaxial cable,** single-conductor kind, whose impedance is seventy-five ohms.

LEAF SHUTTER. *See* **between-the-lens shutter.**

LEARNING CENTER.
A means of providing independent learning experiences in the classroom whereby several stations (each a learning center) are set up in the classroom. At each station the learning tasks are laid out along with instructions for completing them. Usually, all of the necessary materials and equipment are to be found at the learning station, or students are told where to find them. The student's objectives are clearly specified and a means of self-evaluation is provided for each. Learning centers can also be set up in the **media center** of the school, usually because materials and equipment are readily available there. Sometimes the media center itself is termed the learning center.

LEARNING RESOURCES CENTER. *See* **Media Center.**

LENS, PROJECTION.

Projection lenses for **motion picture projectors, slide projectors,** and **filmstrip** projectors come in several different **focal lengths**. They are used to sharply define or focus the **image** on the **screen**. The type of lens to use depends on the size of the projected material and the distance of the projector from the screen. The focal length of the projection lens determines the size of the screen image obtained at a given projection distance. The shorter the focal length, the larger the screen image at a given distance. However, the picture size on the screen can be adjusted by changing the distance of the projector from the screen. Moving it toward the screen will produce a smaller and brighter image—moving it away produces a larger but less bright image. Generally, the 16mm **film** projectors used in schools are equipped with a two-inch focal length lens and are adequate for classroom use. A highly desirable feature is a **zoom lens**, which allows the picture size to be adjusted without changing the position of the projector.

LENS SPEED.

The amount of light that can pass through the lens varies greatly depending on the lens design. A lens that allows more light to pass through can properly expose the **film** more quickly. Therefore, the lens that passes more light is said to have a faster lens speed.

Lens speed is usually specified as an f/number (or f/rating). This f/number is obtained by dividing the diameter of the lens opening into the **focal length** (when focused at infinity) of the lens. Thus, a one-inch diameter lens opening which is one inch from the film (when focused at infinity) would have a lens speed of f/1. The lower the f/number, the faster the lens speed. Setting a lens to a particular f/number is ordinarily described as setting the f/stop.

LENS STOP.

The opening in the lens through which light can pass and which is determined by the **diaphragm**. Lens stops are usually expressed by a given set of numbers. The numbers are derived by dividing the **focal length** of the lens by its effective **aperture**. The resultant number (i.e., lens stop) is expressed by preceding it with a lower case f usually followed by a virgule. Examples of common lens stops are f/2.8, f/5.6, f/11, and f/16. A common rule is, the lower the number the narrower the **depth of field**, and vice versa. Therefore, a lens stop of f/16 will have a greater depth of field than f/2.8 and more of the subject photographed at f/16 will be in focus. One other rule to remember, the higher the lens stop number, the smaller the aperture. Therefore, a compensatory decrease in **shutter speed** is usually required.

LENS TURRET. *See* **turret, lens**.

LENTICULAR SCREEN. *See* **screen**.

LEVEL.

A term commonly used to mean the **amplitude** of a **signal**. For instance, in **audio** equipment, a level control acts to increase or decrease the **decibel** value (or volume). A **videotape recorder** may have a **video** level control that acts upon the amplitude of the video signal; its effect is perceived as a change in the brightness of the picture.

LEVEL INDICATOR. *See* **VU meter**.

LIBRARY, COMPUTER.
A collection of **program**s, **data file**s (be they on **magnetic tape**, **disk**, etc.), and, occasionally, appropriate **documentation**. Such a library is almost always organized in order to be as useful as possible and not necessarily along the line of a traditional library of books or other printed materials.

LIFT PROCESS (PICTURE TRANSFER). *See* **transparency**.

LIGHT-EMITTING-DIODE (LED).
A two-terminal semiconductor electronic device which, when excited by an electrical current, emits light (usually red in color). Often LED or **fluorescent light** displays are used as **signal strength meter**s in **audio** equipment.

LIGHT METER. *See* **exposure meter**.

LIGHT PEN.
A hand-held device, shaped rather like a pen, whose tip contains a photosensitive device or **photoelectric cell**. When the tip is pressed against the viewing surface of a **cathode-ray tube**, measurements can be made from the **screen** or the **terminal** display can be altered. A light pen may also incorporate a light source (e.g., a **light-emitting-diode**) as well as a light sensor, and, in such a combination, may be used to **read** those omnipresent **bar code**s.

LIGHT TRAP.
Any means to prevent light from entering a room (or **film** container) in which such light may cause the premature **exposure** of photosensitive materials.

LINE COPY.
A document whose background is one color (usually white) and whose printed **information** is conveyed in a different, ordinarily contrasting color, most often black. In line copy there is no use made of **halftone**s, etc.

LINE CORD.
A two- or three-wire cord, sometimes called the power cord, that connects an electrical device to the electrical power supply. The three-wire cord, which terminates in a three-prong **plug**, is grounded and protects against electric shock when it is connected to a properly grounded three-socket **jack**.

LINE INPUT.
A **jack** on a **magnetic recording** device that allows the device to accept the line **output** (called the **tape out**) of an **amplifier**, etc., as opposed to a **microphone input**.
A line input will usually accept from 120 thousandths of a volt to 3 volts, while a microphone input will accept only from about 1.2 thousandths of a volt to 30 millionths of a volt.

LINE NOISE.
Noise which occurs in a communications line and which may cause a loss of **data** or **information**. Line noise is frequently caused by the communications system itself.

LINE PRINTER.
A high speed **printer** which prints a full line of copy at a time, rather than one **character** after another, sequentially. Even faster is the **page printer**.

LINE TIME. *See* **connect time**.

LINEAR PROGRAM. *See* **programmed instruction**.

LIQUID CRYSTAL DISPLAY.
A **data** display, usually a **digital display** which, rather than using glowing **light-emitting-diode**s, uses a liquid crystal material encased between two transparent sheets (usually of glass). The viewing side is etched with **character** forming masks and, when an electrical current is passed through the appropriate electrodes, characters (usually black or brown in appearance) form on the viewing surface. Although a liquid crystal display emits no light, it is easily viewed in relatively low light levels, and it has the advantage of requiring less energy than a light-emitting-diode display.

LIQUID TONER. *See* **toner**.

LISTENING CENTER.
A general term that refers to various setups or systems in which the **instructional** method emphasizes listening. Responses from the student may or may not be required. A **media center** is, to a certain degree, a listening center. Language laboratories, which are used for the purpose of teaching foreign languages, are listening centers. Special listening centers may be set up in the classroom so that by means of **headphone**s, students may listen to **program**s originating on **tape recorder**s, **phonograph**s, etc. The listening center may also be one of the learning stations in a series as defined by the term **learning center**.

LOAD.
To feed **data** or **program**s into **storage** or **memory**.

LOGIC CIRCUIT.
A circuit which, of late, is being incorporated into more and more **tape recorder**s. It prevents such mishaps as tape spillage or breakage caused by misuse of the controls of the recorder. It sorts out the commands the user gives the recorder by activating various control switches, and it allows the tape recorder to react in proper time and sequence only to the last of these commands.

LOG-OFF. *See* **protocol**.

LOG-ON. *See* **protocol**.

LONG SHOT.
A picture taken by a camera that is, or appears to be, at a considerable distance from the subject. A human figure would occupy less than one-third the height of the picture. Opposite of **close-up**.

LOUDNESS.
A function of **amplitude**. As a **signal**'s amplitude is increased and fed to a **loudspeaker**, if it is audible, its progressive amplitude increase causes the signal to get louder. The same is true with a visual signal. Increases in amplitude create the visible effect of an increase in brightness, which is the visual correspondent of loudness.

LOUDNESS COMPENSATION. *See* **Fletcher-Munson curves.**

LOUDNESS CONTROL. *See* **Fletcher-Munson curves.**

LOUDSPEAKER.
This is sometimes simply called a speaker, but it is properly an electroacoustic **transducer**, a term that is both a functional and a descriptive designation, in that the loudspeaker converts electrical impulses into acoustic impulses. It changes the electrical **output signal** of the **amplifier** and makes it audible.

There are two basic types of loudspeaker: the **piston speaker** and the **electrostatic speaker**.

When more than one loudspeaker is housed in the same **enclosure**, or when a number of loudspeakers are used simultaneously and fed one **signal channel**, they are described as a loudspeaker system. Ordinarily, the kinds of loudspeakers used in a system are the **midrange driver**, the **tweeter**, and the **woofer**. These are all piston speakers.

LOW-EFFICIENCY.
This describes a type of **loudspeaker** system which requires comparatively more electrical power to create sound, as distinguished from **high-efficiency** type systems, like the bass reflex type (*see* **baffle**).

Most low-efficiency systems tend to be of the **acoustic suspension** design. Generally, a low-efficiency system is one that requires twenty-five or thirty watts or more to produce room-filling sound, while a high-efficiency one can do this with ten or fifteen watts.

Efficiency is not a measure of quality—it is merely descriptive of power requirements and consumption.

LOW-FILTER SWITCH. *See* **filter switch.**

LOW LEVEL. *See* **level.**

LOW-LEVEL LANGUAGE.
A language used to write a **computer program** which is more machine or procedure oriented than a **high-level language**. In a low-level language, each instruction has an equivalent in machine **code**.

LUMEN.
A unit for measuring the flow of light. It is equal to the amount of flow through a unit solid angle from a uniform point source of one international candle. The international candle is a unit of measure of the intensity of light, equal to the light given off by the flame of a sperm candle seven-eights of an inch in diameter burning at the rate of 7.776 grams per hour.

MHz. *See* **megahertz.**

MIS. *See* **management information system.**

mm. *See* **film.**

MACHINE LANGUAGE.
A language, or set of instructions comprehensible to a **computer** and similar to a **low-level language**. In effect, it is the format of a **program** and is immediately capable of being executed by the device concerned.

MAGIC EYE.
A miniature **cathode-ray tube** which used to be used in **tuners** and **receivers** to indicate accurate tuning of the desired station.

MAGNETIC BOARD. *See* **chalkboard**.

MAGNETIC CARTRIDGE.
This is one of two basic kinds of **phono cartridge**; the other kind is known as the **piezoelectric phono cartridge**. There are three types of magnetic cartridges:
1. Moving coil cartridge, in which two tiny coils of wire are physically moved by the **stylus** cantilever (within a magnetic field) and thus have an electrical current induced in them. The advantage of this type of magnetic cartridge is that it has a lower moving mass (but also a lower **output signal** voltage) than the moving magnet type.
2. Moving magnet type. Here the magnets are moved by the stylus cantilever, between a pair of stationary coils, again inducing a voltage in the coils.
3. The induced magnet cartridge utilizes both fixed magnets and coils. Two tiny metal plates are activated by the stylus cantilever to induce an electrical voltage in the coils. This type of magnetic cartridge is not widely used.

Magnetic cartridges ordinarily produce an output signal voltage from about one one-thousandth to ten one-thousandths of a volt. All things considered, the **frequency response, trackability, compliance**, durability, and various measures of **distortion** tend to be better in the magnetic cartridge than in the piezoelectric cartridge. This is particularly true for **stereophonic** and **quadraphonic phonograph** recordings.

The frequency response of a good magnetic cartridge for stereophonic use should be at least forty **hertz** to twenty thousand hertz plus or minus two **decibels**. Its **tracking force** should be no more than two grams, and less, if possible. **Channel separation** should be at least minus thirty decibels. **Crosstalk**, therefore, should be no less than thirty decibels.

MAGNETIC CHALKBOARD. *See* **chalkboard**.

MAGNETIC CORE. *See* **core**.

MAGNETIC DISK. *See* **disk**.

MAGNETIC DRUM. *See* **drum**.

MAGNETIC RECORDING.
A system of recording **audio** and visual material upon a magnetizable medium. Historically, the practice of such recording began in 1898 with the Danish inventor Valdemar Poulsen. In his system, a magnetic steel wire was the medium. Ultimately, a flat **magnetic tape** replaced the unwieldy wire. The first commercially made **videotape recorder** was produced in 1956. In this system both audio and visual material could be recorded on a magnetic tape.

The components of a tape recorder include the **tape transport, head**s, and **recording amplifier**. For **videotape** there is the addition of a **video head**, usually a **helical head**.

The advantages of audio tape recording are numerous: there is no need for processing before playing back; the sound quality (with good equipment) can be of the best; the equipment is usually portable and relatively simple to use; and in **stereophonic** and **quadraphonic** recording, the **channel separation** is considerably better than it is in **phonograph** discs.

MAGNETIC RECORDING (cont'd)

The advantages of videotape recording are essentially two: no processing is required (as distinct from motion-picture **film**), and the operator can **erase** a previous **program** and re-use the tape. Both of these features are shared with audio tape recording.

The future of **digital recording** bodes well for even better quality magnetic recording.

MAGNETIC RECORDING TAPE TYPE. *See* **tape type**.

MAGNETIC SOUND TRACK. *See* **stripe**.

MAGNETIC TAPE.

This is used in **audio, computer,** and **video** type **magnetic recording**. For audio use, the tape is ¼-inch wide—except in the audio **cassette** where it is slightly less than ⅛-inch wide—and it varies in the thickness of its **base** from ½ **mil** (1 mil equals .0001 inch) to 1.5 mils. **Videotape** may vary in width from ¼-inch to 2 inches and may be from ½ mil to 2 mils in thickness. Besides the base, magnetic tape is made up of a **coating** on the base. The formulation of the coating varies. Ferric oxide is the most usual formulation for videotape. Audio tape may be any of four **tape type**s.

The coating is composed of microscopic particles of a magnetized material—the smaller these particles (and the more uniform in size and evenly dispersed and oriented they are), the better the quality of the recording will be.

All magnetic tape should be stored in individual containers, preferably vertically, and at normal temperatures (sixty-to-eighty degrees, preferably). Furthermore, to avoid inadvertently erasing the **program**, the tape should not be stored near magnetic fields, such as those generated by **loudspeaker**s.

Mini- and larger computers use "tape drives" (computer **tape recorder**s) designed specifically for computer use, and use ½-inch wide tape, most often in 2400-foot lengths and mounted on **reels**. Home and other **microcomputer**s often use audio tape cassettes and recorders.

MAGNETIC TAPE TYPE. *See* **tape type**.

MAGNETIC TRACK. *See* **stripe**.

MAGNIFICATION RANGE. *See* **diameter**.

MAIN FRAME.

A large **computer,** that is, one larger than a **microcomputer** or a **minicomputer**. At one time, main frame referred to the **central processing unit** of a computer, but that was prior to the widespread use of the smaller devices (i.e., mini- and microcomputers).

MAIN STORAGE.

The central **storage** unit of a **computer,** as compared with **auxiliary storage**. Main storage can usually be accessed directly, while auxiliary storage can only be accessed indirectly, i.e., after it has somehow been "plugged in" either manually or by the entry of certain commands which call it into play.

MAINTENANCE.

Any activity whose intent is to keep **data, hardware,** or **software** in proper working condition. Preventive maintenance is done regularly to preclude anticipated failures. Corrective maintenance is work done to fix an already existing problem. File maintenance

110 / MAINTENANCE

MAINTENANCE (cont'd)
is the deletion, addition or other correction to a **computer** file (which may contain data or software).

MANAGEMENT INFORMATION SYSTEM (MIS).
Usually a **computer information** system (as opposed to a manual one) designed to aid managers in making their decisions. Such a system usually contains a large number of **data** concerning a variety of factors related to the organization being served.

MASK (PHOTOGRAPHY).
Either an opaque material designed to prevent **exposure** of specific areas, usually the border of a **print**, or any material (frequently an adhesive-backed, colored tape) used to act as the border of an **image**, etc.

MASS STORAGE. *See* **storage**.

MASTER CONTROL.
A control that raises or lowers the **level** of the **signal** (in either **audio** or visual devices) and overrides the other, individual, level controls. For instance, a **quadraphonic** audio **amplifier** may have four level controls—one for each **channel**, plus a master control. The individual controls are used to achieve **balance**; the master control affects all four channel levels proportionally, while maintaining the balance.

MATRIX FOUR CHANNEL.
A method of recording a **quadraphonic** type **program** on a **phonograph disc**. The primary types of matrix four **channel** approaches are the SQ (developed by Columbia Records) and the QS, or Vario Matrix (developed by the Sansui Electronics Corporation). Both of these approaches employ "logic" in the **decoders**, which separate the four channels of **information** from the two channels engraved in the single groove. The purpose of the so-called logic, as used in these decoders, is to increase the **channel separation**.

There are now **amplifier**s, **preamplifier**s, and **receiver**s that incorporate matrix four channel decoders, and decoders are also available separately.

In competition with the matrix four channel system was the system developed by RCA, called **CD-4**. Unless the matrix four channel decoder makes use of various types of logic, there tends to be a loss of rear left-to-right channel separation; the CD-4 approach seemed to suffer less from this type of quadraphonic denigration. CD-4 **software** is no longer being produced.

Matrix four channel is sometimes referred to as "4-2-4" system.

MATTE.
A surface of low **reflectance**, i.e., dull (opposite of **glossy**) and often used to refer to a type of **print** paper. Matte surface prints are sometimes called European style or "studio" prints. Matte print papers are available for either black-and-white or color prints.

MATTE WHITE SCREEN. *See* **screen**.

MECHANICAL CONTROL.
The control on a device (e.g., the **fast forward** switch on a **tape recorder**) which acts through connecting mechanical linkages. A mechanical control requires greater effort, by the operator, to effect, and is subject to greater wear (and ultimate failure) than is a **solenoid** or **solid state** control. Another disadvantage of the mechanical control is that,

MECHANICAL CONTROL (cont'd)

prior to going from one **mode** of operation to another, the user must usually stop the device. This is not necessarily the case with solenoid controls.

MEDIA CENTER.

Variously called an instructional materials center, learning resources center, or **learning center**, this is a place (section, room, or building) where all sorts of audiovisual materials and devices are located and made accessible to the student or teacher. The collections can be centralized in one place or decentralized throughout the campus. A media center may include the library or be separate from it. It may include study **carrel**s with various **multimedia** capacities, including television and **computer assisted instruction**. Various multimedia packages or kits, which include printed materials, **magnetic tape**s, **slide**s, **film**s, etc., are cataloged, stored, and made available for student use in the media center.

MEDIUM.

The singular of *media* and possessed of a variety of meanings. Usually, medium is understood to be the "means" of conveying **information**. For example, **magnetic tape**, **punched card**s, newspapers, **film**, **phonograph disc**s, **cathode-ray tube** viewing surfaces, etc., can all be considered media. However, means of communication, like telephone lines, television broadcasting, etc., are also construed to be media. Since the present-day library contains many of these media, and since books and other printed materials are also media, it is no wonder that many libraries prefer to call themselves **media centers**.

MEGAHERTZ (MHz).

A combination of the multiplication factor one million, plus the term **hertz**; for example, five megahertz equals five million hertz.

MEMORY.

Usually synonymous with **main storage**, memory may also mean any **storage**, whether main or **auxiliary storage**. At present, memory is usually composed of **bubble memory**, **disk**, **magnetic tape**, **random access memory** or **read only memory** chips. It should be remembered that memory can only store **digital data** (or **bit**s) and is not like human memory wherein leaps of imagination can be made or "real intelligence" stored.

The size of memory is usually measured in thousands of **byte**s, e.g., 100K bytes equalling 100,000 bytes.

MENU.

A display listing, usually on a **cathode-ray tube**, of the variety of capabilities in **storage** from which the human operator may select his desired (or required) **program**s, **data**, etc.

METAL TAPE.

A **magnetic recording** tape **coating** which was introduced in 1979. Like **chromium dioxide tape**, metal tape (or sometimes, metal-particle tape) theoretically can improve the **dynamic range** and, particularly, the recording of higher **frequency** (or **treble**) sounds. At the time of this writing, although most modern audio **cassette deck**s can record and play back metal tape, the cost of the tape itself is sufficiently high as not to justify the improved sound the tape may produce.

MICROCARD.

A **microform** which is obsolescent. Its form is an opaque card (usually 3x5-inches), and it requires special devices to achieve sufficient magnification for reading.

MICROCOMPUTER.

A small, usually desk-top-sized, **computer** whose **central processing unit** may consist of a single **integrated circuit** type **chip**. Most microcomputers include a **keyboard terminal** and **disk drives** for **floppy disks**. Many may incorporate a **cathode-ray tube** display and **printer** of some sort. Many microcomputers use **operating system**s and **program**s that can be plugged into the system. Microcomputers usually offer relatively limited **memory** (compared to a **minicomputer** or **main frame**), usually limited to something less than 600,000 **byte**s.

MICROFICHE.

A French term that means miniature index card. It describes a 4x6-inch **microform** which ordinarily contains approximately ninety-eight transparent miniaturized pages. Microfiche can be read as projected enlargements on a microfiche **reader screen** or printed on a paper enlargement, usually 8½x11 inches. The great advantage of this system is that it can provide users with publications to which they might not otherwise have access.

MICROFICHE GRID.

The array, or pattern, of **image**s or **frame**s on a sheet of **microfiche**. In a standard microfiche, the grid consists of a total of ninety-eight frames, arranged in seven horizontal rows, and fourteen vertical files. If the grid is not exactly arranged on the sheet, automatic **blowback** cannot be achieved, nor is it easy to move from frame to frame on a **reader** or **reader/printer**.

MICROFILM.

Films of either 35mm or 16mm width on which documents, printed pages, etc., are photographed in a reduced size. The process of microfilming involves photographing only one or two pages of a book or document on a single **frame**. There are several types of microfilm **reader**s available on the market which make it possible to read the projected **image**s directly on a **screen** or to have enlarged copies produced as paper prints or transparencies. Some microfilm readers are capable of handling all three **microform**s: microfilm, **microfiche**, and **microcard**s.

MICROFILM READER. *See* **reader**.

MICROFORM.

The general term for various types of **information storage** media that maximize efficiency of storing and retrieving of printed materials, documents, pictures, etc., by miniaturizing them through photography and using the **film** or derivatives from it. Examples: **microfilm, microfiche, microcard,** and **photo-chromic-micro-image**.

MICROGRAPH.

The reproduction of the **image** of an object seen through a microscope. The term *micrographics* has achieved some currency in a different sense, that being the creation or reproduction of **microform**s. A more proper term is *microreproduction*.

MICROGRAPHICS. *See* **micrograph**.

MICRO-OPAQUE. *See* **microcard**.

MICROPHONE.

An electroacoustic **transducer** which converts sound waves into electrical impulses. There are two basic types of microphone: the **dynamic microphone** and the **condenser microphone**. Either type may be designed as a **cardioid pattern** or a more or less **omnidirectional microphone**.

Microphones come in a variety of sizes and shapes with a number of accessories. Some of the more important adjuncts to a microphone are a pop screen (which helps filter out the "pops" that occur when the letters p and b are pronounced too closely to the microphone), a wind screen (for outdoor recording), and stands (available as desk stands, floor stands, **boom**s, etc.).

Two important **output** measures of a microphone are its output voltage (usually between 1.2 millionths and 30 millionths of a volt) and **impedance** (high impedance ranges from one thousand to twenty-five thousand ohms; low impedance ranges from twenty-five to six hundred ohms). Better microphones tend to be of low impedance type; as impedance rises, there is a concomitant high **frequency** loss.

Microphones that are to be used only for recording or broadcasting speech need have a **frequency response** of from one hundred or two hundred **hertz** to six thousand or eight thousand hertz plus or minus four **decibels**. For use with music **program** sources, the frequency response should be at least sixty hertz to fourteen thousand hertz plus or minus three decibels.

MICROPHONE BOOM. *See* **boom microphone**.

MICROPHONE INPUT.

A **jack** allowing **input** on a device like an **amplifier** or **tape recorder** whose purpose is to accept the **plug** of a **microphone**. Its design allows it to handle from about 1.2 thousandths to 30 millionths of a volt. This is different from the **line input**, which usually accepts from 120 thousandths of a volt to 3 volts.

MICROPHOTOGRAPH.

A photograph which is so small that it requires magnification (e.g., a microscope, etc.) to be viewed.

MICROPRINT.

Printed or written matter which, by means of microphotography, has been so greatly reduced in size that it can be read only through a magnifying device. The usual **medium** for microprint is paper. An example of this format is *The Compact Edition of the Oxford English Dictionary* (London: Oxford University Press, 1971).

MICROPROJECTION.

A means by which an entire class (especially in the biological sciences) can be shown the **image** in a microscope at the same time. Microprojectors are either attachments for a microscope or separate projectors with built-in microscopic lenses. They are used in much the same way as **slide projector**s are, but their adjustment is more complex.

MICROPUBLICATION.

As a verb, the act of publishing any previously unpublished material, but in a **microform** format. As a noun, the microform so published.

114 / MICROPUBLICATION

MICROPUBLICATION (cont'd)
A fine distinction is made between micropublication and microrepublication, this latter being previously published (regardless of whether in **hard copy** or microform format) materials which are republished in microform format.

MICROREPRODUCTION. *See* **micrograph**.

MICROREPUBLICATION. *See* **micropublication**.

MICROSCOPE SLIDES. *See* **photomicrography**.

MID-RANGE DRIVER.
A component of many **loudspeaker** systems which is, in itself, a loudspeaker, but one of relatively restricted **frequency range**. Usually, the mid-range **driver** reproduces frequencies from about five hundred **hertz** to about three thousand hertz. Frequencies above and below this range are ordinarily handled by "drivers" termed the **tweeter** and **woofer**, respectively. The mid-range driver is occasionally referred to as the squawker.

MIL.
The measure applied to **magnetic tape base** thickness. One mil is equal to .001 inch.

MINICOMPUTER.
A **computer** larger than a **microcomputer** but smaller than a **main frame**. It usually has a **memory** in excess of 200,000 **byte**s, and has a more sophisticated complement of capabilities such as its **operating system** (which may be on **disk**), **auxiliary storage** (often on **magnetic tape**), and may serve more than one **terminal** for **input/output**.

MIXER.
A device that can be a separate unit or can be built into a **tape recorder**. The mixer allows the user to blend and **balance** more than one **input** for **audio**. It can be used so as to allow only the blending of **microphone input**s; for example, four separate **microphone**s could be connected to a **stereophonic** type **open-reel** tape recorder—one microphone each, say, for two narrators and the other two microphones used to capture the sounds of a group of musicians.
Another type of mixer can accept both microphone inputs and **line input**s. This type allows blending the **signal**s from a pair of microphones (narrators, perhaps) and a set of stereophonic music signals from a **phonograph**, another tape recorder, etc., in order to provide a balanced musical background. As stated previously, many of the newer tape recorders have mixers (usually of the latter variety) built into them to provide greater recording flexibility.

MODE.
In **audio** or **video** equipment this usually refers to the manner of audio presentation, whether **quadraphonic, stereophonic, monaural,** etc. When applied to **tape recorder**s, mode often means the operational function, i.e., the record mode or playback mode.

MODE SELECTOR.
In **audio** equipment this is the switch that allows the user to select the **mode** of audio presentation, whether **stereophonic, monaural,** etc. In a **tape recorder**, it may also refer to the switch(es) that cause the machine to operate in either the record mode or the playback mode.

MODEM.

An **interface** device between a **terminal** (or **computer**) and a communications line, which usually connects to another terminal or computer. The modem, in effect, "translates" the **data** generated by the **output** device into electrical impulses compatible with telephone transmission. At the receiving end, another modem is usually required to "retranslate" the impulses back into data. Modem is synonymous with **data set**. There are some telephones which incorporate a modem into their construction: these are called modemphones. The modemphone is almost synonymous with the **data-phone**. Modem is a contraction of "modulator-demodulator." The device modulates **digital signal**s into telephone line **analog** signals when transmitting and does the inverse when receiving. The acoustic coupler is a modem which connects a telephone to a terminal or computer.

MODEMPHONE. See **modem**.

MODULAR PHONE CONNECTOR.

A new type of connector used by the Bell Telephone Company (and most telephone manufacturers) to join the handset of a telephone to the base instrument, and the base instrument to the telephone line (usually terminating in a wall **jack**). The modular connector enables the user to change, or move (if wall jacks are installed) telephones without the aid of a service person.

MODULATION.

A method of both broadcasting and recording in which the broadcast or recording **information** is altered at the point of broadcast (or in the recording) so that, in effect, it is compressed. The information is later sorted out within the devices that receive the broadcast or that play back the recording. In this way, for example, a radio station may broadcast in **stereophonic** a **frequency modulation** (FM) band **program** or a recording company may encode **quadraphonic** material on the two sides of the groove in a **phonograph** disc. In these instances, the proper equipment will "demodulate" the information and provide the material either stereophonically or quadraphonically.

MODULE.

A discrete unit of a **program** and which can be identified by its function (e.g., compiling or **editing**).

MONAURAL.

Although this literally means one ear, it is popularly, but improperly, used to mean "monophonic." In this sense it refers to **audio** information contained in one **channel**, as opposed to **stereophonic** or **quadraphonic**. As an example, a simple **amplitude modulation** (AM) radio is almost always monaural in that it receives only one channel and has but one **amplifier, loudspeaker**, etc.

MONAURAL DISC. See **disc**.

MONITOR.

Used as a noun, this is either the device or the circuit that allows the recordist to hear (or watch) what is being recorded as the recording process is going on. As a verb, monitor has come to mean this process of hearing or watching.

The user can monitor a **tape recorder** in either or both of two ways. In the first, which is usually termed **source monitoring**, the listener sees and/or hears the **program** being recorded as it is seen and/or heard by the **recording amplifier**. Although this is better than not being able to monitor at all, it is considerably less useful than the second type of

MONITOR (cont'd)

monitoring, called **tape monitoring**. Here, what actually has been recorded on the tape is seen and/or heard a fraction of a second after recording. In order to offer tape monitoring, the recorder must have a separate **record head** and **playback head**. Most such machines are described as having three heads (the third being the **erase head**). Further, in order to utilize the monitoring ability of a tape recorder (assuming that it is used in the context of a larger system), the **amplifier** or **preamplifier** to which it is connected must have a "monitor" switch.

In a **videotape recorder,** the same principle obtains, both for the **audio** portion and for the picture—thus allowing either source or tape monitoring of the audio and the **video**.

The whole rationale for monitoring is to allow the user to correct any problems as the recording is being made.

Monitor also refers to a television set especially designed to accept the **input**s from a **videotape** or **video cassette** device (or from an **educational television, closed circuit** system). Such monitors may or may not contain a **tuner**, which enables the device to receive regularly broadcast television programs.

MONOPHONIC. See **monaural**.

MONOPOD (PHOTOGRAPHY).

As compared with the usual camera stand (i.e., **tripod**), a "one-legged" stand on the top of which is a mount to which the camera can be secured and which aids in steadying the camera during shooting. Although not as sturdy or stable as a tripod, the monopod is better for cinematographic work or slow **shutter** speeds than is hand-holding the camera.

MOTION PICTURE FILM. See **film**.

MOTION PICTURE PROJECTORS.

Different types of projectors are designed for the various **film** sizes, but schools generally use 16mm and 8mm projectors. A sound motion picture projector basically consists of an intermittent mechanism (**pawl-sprocket** or claw) which is driven by an electric motor. This moves the film past a **gate** and an **aperture**, illuminated by a **projection lamp**. The lighted **image** is then projected onto the **screen** via the lens. Most projectors have a **framing** device which, when adjusted, permits only one complete picture to be seen on the screen. Also, the lens is adjusted to focus the image.

Film is fed through the film channel to the gate by a **supply reel**, and after it passes the sound system it is wound up on a **take-up reel**. At a suitable point past the gate and while the film is in constant motion, it passes over either a sound pickup **head** or a sound drum, depending on whether the projector is designed for film with a magnetic sound **stripe** or an **optical sound track**. The magnetic head or sound drum is connected through **amplifier**s to a **loudspeaker** system. The speaker can be mounted in the projector or it can be a separate unit. **Volume** and tone are adjusted by the sound control knob(s).

If a film has a magnetic sound stripe, the heads in the projector (like those of a **tape recorder**) play back or record sound from a magnetic stripe on the film. If the film has an optical sound track, light from an **exciter lamp** in the projector passes through the film sound track to the sound drum, where a **photoelectric cell** converts the pulsing light into electrical impulses.

Threading: Some 16mm projectors have to be threaded by hand. That is, in order to put the film into the machine, the **leader** has to be manually guided through the film path of the projector and inserted in the take-up reel. Usually each type of projector has a threading diagram of some sort. To start with, one should be sure that there are at least five

MOTION PICTURE PROJECTORS (cont'd)

feet of leader to thread into the projector, and also that the take-up reel is as large as (or larger than) the supply reel. The leader (or film) must be fitted snugly over the teeth in each **sprocket** wheel. Most importantly, film loops should be established above (and below, if called for) the film channel at the gate. The film must lie flat in the film channel of the gate so that the claw or pawl-sprocket can properly engage the perforations in the film. When the film (with optical sound track) is passed around the sound drum, it must be snug and under tension, because if it is too loose the sound will be distorted. To complete the threading, the film end is inserted in the take-up reel, which should be turned once or twice in a clockwise direction.

An automatic threading or autothreading feature is included on some projectors. When the projector controls are properly set, the film is simply inserted in the film slot and the machine is turned on. It automatically performs all remaining threading operations except for insertion of the film end in the take-up reel.

Generally, 8mm projectors are smaller than 16mm machines, but they operate on the same principles. For operation and threading (if this is necessary), it is advisable to refer to manufacturers' manuals for specific models. The Kodak Ektographic Sound 8mm projector provides automatic threading through the gate and sound mechanism, but the film has to be manually attached to the take-up reel. It plays back sound from a magnetic sound track on the film, but it cannot record sound. Cartridge-loading Super 8mm film projectors need only have the **film cartridge** inserted in the machine; no further manual threading is necessary.

MOUNTING OF SLIDES. *See* **slide**.

MOVING COIL CARTRIDGE. *See* **magnetic cartridge**.

MOVING MAGNET CARTRIDGE. *See* **magnetic cartridge**.

MULTIMEDIA.

The use of more than one **medium** of audio-visual instruction in a **program** or presentation. For instance, a program that incorporates the visual medium of 35mm **slides** plus an **audio** medium of recorded music would be a multimedia production.

MULTIMEDIA KIT or PACKAGE.

This term can be applied to any of a number of various combinations of audiovisual materials such as printed materials, pictures, records, tapes, **slides**, and **films**. The material can be used by the teacher as part of the instructional **program** in the classroom, but it is usually cataloged, stored, and made available for student use in the **media center**.

MULTIPATH.

The reception of the same **signal** reaching a **receiver** by two or more paths, causing distortion in radio equipment and ghosts in television. Ordinarily the signals are received within fractions of a second of each other and are often caused by bouncing of the signals off buildings, hills, clouds, etc. There are some devices which help suppress multipath reception, but the best cure is a good **antenna**, usually on a mast of sufficient height to clear the object causing the bounce (except for clouds, of course).

MULTIPLE EXPOSURE.
More than one **image** on a **frame** of **film** caused by more than one **exposure** of that frame. If there are but two images, the result is usually termed a double exposure. Although multiple exposures are usually made unintentionally, occasionally experienced photographers consciously employ them for aesthetic effect. Most modern cameras are designed to prevent all but intentional multiple exposure. The most frequent cause of accidental multiple exposure is the inability of the film to advance, either because of being improperly loaded or because of torn **sprocket** holes.

Multiple exposure is occasionally referred to as overlap or image overlap.

MULTIPLEX.
A term used to describe **frequency modulation** (FM) type **stereophonic** radio broadcasting.

MULTIPLEXER (TELEVISION).
A device that uses two mirrors or prisms, in order to allow one TV **film** camera to be used with two or more TV film projectors.

MULTIPROCESSOR.
1. A **computer** installation with more than one independently **programm**ed **central processing unit** sharing **main storage**.
2. A computer installation with one or more central processing units which is capable of processing two or more programs or segments of a program simultaneously.

MUSIC POWER. *See* **IHF power.**

MUTING.
A circuit in a **frequency modulation (FM) tuner** or **receiver** which allows interstation **noise** or weak (or remote) stations to be obliterated by silence. When muting is employed, there appears to be no static or weak stations between the stronger stations on the **band**. Good devices that offer muting usually allow the user to shut the muting circuit off and/or to lessen its effectiveness in order to receive less powerful stations when so desired.

MYLAR TAPE. *See* **polyester tape.**

NAEB. *See* **National Association of Educational Broadcasters.**

NAVA. *See* **National Audio-Visual Association.**

NCAT. *See* **National Center for Audio Tapes.**

NET. *See* **National Educational Television.**

NPR. *See* **National Public Radio.**

NATIONAL ASSOCIATION OF EDUCATIONAL BROADCASTERS (NAEB).
An organization that serves the professional needs of educational radio and television stations, systems, and personnel. Its Educational Television Stations (ETS) division includes a **program** service for **public television** stations. Through cooperative efforts it distributes **frequency modulation** (FM) radio program materials to college- and university-owned stations, those owned by school systems, and privately owned nonprofit stations. It

NATIONAL ASSOCIATION OF EDUCATIONAL BROADCASTERS (NAEB) (cont'd)

also participates in developing rules and regulations for the industry and speaks for the future needs of educational broadcasting. Publications: *Educational Broadcasting Review* (bimonthly) and occasional publications, including *Memos on Instruction*. Address: NAEB, 1346 Connecticut Avenue, N.W., Washington, DC 20036.

NATIONAL AUDIO-VISUAL ASSOCIATION (NAVA).

A national multipurpose association of media **hardware** and media **software** producers and manufacturers, dealers, representatives, and others involved with educational and informational activities, services, and products. Address: The National Audio-Visual Association, 3150 Spring Street, Fairfax, Virginia 22031.

NATIONAL AUDIO-VISUAL CENTER.

A United States government organization that seeks to coordinate the efficient use of federal audiovisual materials (primarily 16mm **film**s and 35mm **filmstrip**s) and to provide **data** concerning sources, costs, and availability. Address: National Audio-Visual Center, c/o National Archives and Record Service, General Services Administration, Washington, DC 20409.

NATIONAL CENTER FOR AUDIO TAPES (NCAT).

An important source of **magnetic tape** recordings, it consists entirely of non-commercial **program**s deemed appropriate for educational purposes. Located at the University of Colorado at Boulder, the center is associated cooperatively with the Association for Educational Communications and Technology (AECT) and the **National Association of Educational Broadcasters** (NAEB). Various agencies throughout the country (such as colleges, universities, state departments of education, as well as other private and public organizations) contribute master tapes to the University of Colorado collection. Here, they are classified, cataloged, and duplicated upon direct order from educational institutions. Further information about the Center's services and resources can be found in its catalog, which can be procured from: The Association for Educational Communications and Technology, 1126 Sixteenth Street, N.W., Washington, DC 20036.

NATIONAL EDUCATION TELEVISION (NET).

An organization that is part of the **Corporation for Public Broadcasting** (CPB). Its purpose is to produce **program**s for viewing on **public television channel**s. The Children's Television Workshop of NET produces *Sesame Street* and *The Electric Company*.

NATIONAL PUBLIC RADIO (NPR).

A loose affiliation of essentially **frequency modulation** (FM) radio stations which are principally located on American college and university campuses. These member stations tie in to NPR headquarters in Washington, DC, for live **network** generated **program**s such as *All Things Considered*. Other (usually taped) programs are shared on a syndicated basis. NPR offers both educational programming and programs of an entertainment nature (classical music, folk music, etc.) plus informative programs like talk shows and interviews.

NPR is supported by federal monies; by nonprofit, industrial, and business grants; and by listener donations.

NATURAL LIGHT.

Light given off by the sun, moon, stars, etc., as opposed to **artificial light**. Natural light, sometimes called available light, is often thought to be the best light for photographic purposes and most **film** is designed for its use.

NEEDLE. *See* **stylus**.

NEGATIVE.
That **film** on which an **exposure** is made and which carries a **latent image** until the negative becomes **processed film**, after which the latent image becomes a **real image**. That real image, however, is a "negative" one, in that dark tones of the picture appear light, and vice versa. The true relationship of light to dark is restored in the **print** made from the negative; the print is called a **positive**.

NEGATIVE IMAGE (TELEVISION).
This refers to a picture **signal** that has a **polarity** which is opposite to normal polarity; it produces a picture in which the white areas appear as black, and vice versa.

NEON LIGHT. *See* **artificial light**.

NETWORK.
A group of radio or television stations connected by ownership or some other type of liaison, which broadcast, for the most part, the same **program**s at the same time.

NETWORK (COMPUTER).
The interconnecting of **computer**s via communication lines in order to transmit **data** among the connected devices.

NETWORK SUPERVISOR. *See* **supervisor**.

NEXUS.
A point of connection or interconnection in a **network**.

NODE.
Any station, **terminal** or end- or midpoint in a **network**, but not necessarily a **nexus**.

NOISE.
An undesirable disturbance in a communication system. For instance **hiss**, **rumble**, and **snow** are all "noise." Noise can be considered the opposite of, and an intrusion upon, the **signal**.

NONGLARE GLASS. *See* **nonreflective glass**.

NONREFLECTIVE GLASS.
Glass whose surface has a **matte** finish and whose **reflectance**, thereby, is considerably reduced.
Nonreflective glass is often used in framing pictures and is useful in holding down pages or leaves to be photographed. Still, when using such glass as a weight, it is wise also to use a **polarizing filter**.

NONREVERSIBLE IMAGE. *See* **nonreversing film**.

NONREVERSING FILM.
Film on which the **image** can not be reversed from a **negative** to a **positive**, or vice versa, from **generation** to generation. An example is **diazo film**.

OCR. *See* **optical character recognition.**

OEM. *See* **original equipment manufacturer.**

OS. *See* **operating system.**

OFF-LINE.
Refers to **hardware** not under the direct control of a **central processing unit.** Usually contrasted with **online.**
It is often preferable to perform certain functions off-line in order to allow the central processing unit to do what it does best or fastest. For instance, an off-line **printer** which can print **data** or **information** from a **disk** is slower and less expensive than **on-line** equipment. Its use allows the central processing unit to go on to other functions.

OMNI-DIRECTIONAL MICROPHONE.
A **microphone** that is equally sensitive and receptive to sounds emanating from all directions—from behind and in front of the device. This is in contrast to a **cardioid pattern** microphone.

ONLINE.
Hardware directly coupled to, and controlled by, a **central processing unit.** Usually contrasted with **off-line.**
An example of proper online use is the obtaining of **data** or **information** which can only be provided online for **real time** purposes.

OPAQUE PROJECTOR.
A machine that uses a high intensity **projection lamp** (usually a one thousand watt **incandescent light**) to shine directly on a flat opaque object or picture. Light is reflected from the picture into a reversing mirror, which sends its light through the lens system to the **screen.** The opaque projector usually has a platen that lowers or accepts materials for projection. Flat pictures or even three-dimensional objects can be inserted into the projector by lowering the platen. After the material is inserted, the platen is raised for projection. A severe limitation of this type of projector is that since light loss due to reflective projection is too great, the projected **image**—even in a completely darkened room—lacks the sharp, brilliant quality of a projected **transparency.** The opaque projector, therefore, cannot be used successfully in a normally lighted classroom. The opaque projector is sometimes called the balopticon.

OPEN CAPTIONED. *See* **captioned.**

OPEN-REEL.
A term used to describe any type of **tape recorder** (including **videotape**) in which the **magnetic tape** is on one **reel** (**supply reel**) and is fed to another reel (**take-up reel**), and neither of these reels is enclosed in any kind of housing such as a **cassette.** Open-reel devices are sometimes called reel-to-reel machines. Open-reel **audio** tape is usually one-quarter-inch in width.
The advantages of open-reel audio tape recorders are:
1. Easy access to the tape for **editing.**
2. Better **frequency response** and **dynamic range** due to both the greater width of the tape employed—one-quarter-inch as opposed to less than one-eighth-inch in audio cassette devices—and the faster **tape speed**s available—15 or 7½ **inches-per-second** compared to

122 / OPEN-REEL

OPEN-REEL (cont'd)

3¾ or 1⅞ inches-per-second for audio **cartridge** and cassette devices, respectively.

3. The potential for **sound-on-sound** or **sound-with-sound** recording, if the device offers independent recording on individual **channel**s.

4. The frequent provision in open-reel machines for **tape monitoring**. Most audio cassette machines provide only **source monitoring**.

The disadvantages of open-reel machines are:

1. The relatively large size of both the device itself and the reels of tape as compared with either audio cassette or audio cartridge devices and **software**.

2. Difficulty of use. Open-reel equipment must be threaded. (Some self-threading models have been available, but these are to be shunned because of the frequent damage they inflict on the tape.) This bothersome labor is dispensed with in both audio cassette and audio cartridge machines.

3. Comparatively high cost. A good open-reel **deck** may cost one thousand dollars, while a very good audio cassette deck is usually half that or less.

Open-reel machines will probably continue to be used for high quality recording, while audio cassette devices will be increasingly used in schools and libraries, where less quality or less flexibility is necessary.

OPERATING SYSTEM (OS).

That **software** which controls the execution, scheduling, etc., of **computer program**s for the entire computer system. Because it controls every operation performed by the computer, the operating system usually resides in **main storage** in order to minimize **access time**.

OPERATION.

A single step or function, often specified by a single **computer program** instruction. An example may be "add," as in "add x and y." Not only should the operation be comprehensible to the computer, but the result of the operation must fall within logical bounds. For instance, "one plus one cannot exceed two nor be less than one."

OPERATIONS RESEARCH.

The quantitative, or mathematical, study of **input** and **output** operations (operations here defined as the positive actions and interactions of humans and machines) in order to optimize either the results or efficiencies of those operations. The findings or results of operations research are often used in **systems analysis**.

OPTICAL CHARACTER RECOGNITION (OCR).

Refers to **hardware** which is capable of identifying printed **character**s and, usually, capable of translating them into **digital** ones, often to be used by a **computer**. Optical character recognition, when fully refined and developed, may prove to be the fastest means of **input** of printed **data**.

OPTICAL-DIGITAL. *See* **optical disk**.

OPTICAL DISK.

A **disk** which stores **data**, not by **magnetic recording**, but by means of **bits** optically encoded on (or just under) its surface. These bits usually take the form of microscopic pits which are caused by the controlled burning action of a laser beam.

There are two basic means of playing back an optical disk: mechanically, or by means of a "reading" laser beam. The mechanical system (patented and licensed by RCA) uses a

OPTICAL DISK (cont'd)

stylus similar to the type utilized in playing back a **phonograph disc**. The laser system (patented and licensed by Philips Lamp) has no counterpart.

The data which can be encoded on an optical disk can be purely **digital** or may be in the form of sound and pictures. In the latter, the device is connected to a television set and, when operated, feeds both **audio** and **video** signals to it, just as if they were received through **cable television** or by broadcast means.

The laser-based system can provide a **stereophonic** audio **signal**, the mechanical system but a **monophonic** one.

Each side of a twelve-inch disk can store over fifty thousand **frames** or millions of bits of data.

Although at present neither system allows the user to record his own material, the laser system (perforce there being no mechanical contact with the disk) seems better in that the **software** ought to last indefinitely. The use of the optical disk for data **storage** has not yet come into widespread or practical use.

OPTICAL IMAGE. *See* **image**.

OPTICAL SCANNER.

A specific piece of **hardware** which is capable of providing **optical character recognition** by quickly (and, often, automatically) scanning the printed material.

OPTICAL SOUND TRACK.

A photographic **image** of sound on **film**, produced by a beam of light whose intensity has been caused to vary rapidly because of the electric **signals** from a **microphone**. It varies in its light and shade along the length of the film and represents a record of the sound the microphone received. When the film is projected it moves past a narrow slit illuminated by an **exciter lamp**. The light and shade of the sound track cause the intensity of the light passing through the slit to vary; when this varying light falls on the **photoelectric cell** of the projector, it gives rise to electrical signals that can be amplified and then transmitted to the **loudspeaker**. Another type of sound track is that afforded by the magnetic **stripe**.

OPTICAL VIEWING SYSTEM.

The **viewing system** of a camera in which the user views the subject to be photographed through a lens system separate from the one that will actually take the picture. Since these two lens systems are physically separated on the camera, a slight difference occurs between the **image** seen in the **view finder** and the image projected by the "taking" lens on to the **film**. This difference, or discrepancy, occurs no matter how slight the distance between the two lens systems, and is technically known as **parallax**. Naturally, the closer the two lens systems, the less the parallax effect.

The parallax effect is greater when the subject is relatively close to the camera. With greater distances, the parallax effect is diminished. Most optical viewing system view finders incorporate some kind of marking to indicate, when the subject is somewhat close to the camera, what compensation should be made to ameliorate the parallax.

A system that does away with parallax is the **single lens reflex camera**.

ORIGINAL EQUIPMENT MANUFACTURER.

The organization which makes specific "brand" equipment (or replacement parts therefor), as opposed to one which makes equivalent or compatible equipment or parts for the stated original equipment.

ORTHICON IMAGE CAMERA TUBE.

A television camera **pickup tube** in which the optical **image** falls on a **photoemissive** cathode which emits electrons that are focused on a target at high velocity. The target is scanned from the rear by a low-velocity electron beam. Return beam **modulation** is amplified by an electron multiplier to form an overall light sensitive device. The orthicon image camera is quite expensive because of its complex circuitry and the usual addition of a **view finder**. However, because it has better light-response characteristics than the **vidicon camera tube** and because of the need for on-site operator control, it is widely used in commercial, educational, and **public television**. With special adaptation, both orthicon image and vidicon models are capable of a resolving power up to one thousand lines. In view of cost considerations, it is perfectly feasible for school districts to utilize vidicon cameras with view finders in their **educational television** programs, but this is done at the expense of some flexibility and quality.

ORTHOCHROMATIC FILM (ORTHO).

A special black-and-white **film** whose **emulsion** is not sensitive to red and, therefore, renders red-colored objects dark. A more generally useful film is **panchromatic film (pan)**. Orthochromatic film is principally used in copying work.

OUTPUT.

1. The **signal**(s) emanating from an electronic device, as opposed to those which are fed to it (**input**). One device's output, therefore, becomes another's input—i.e., the output from an **amplifier** is the input to a **loudspeaker** to which the amplifier is connected. When such devices are connected, aspects such as **impedance**, voltage, etc., must be matched so that the output of one is acceptable as input to another.

2. **Computer** output is **data** or **information** transferred from **storage** to **hardware** such as **terminals** or **printers**. It is the opposite of **input**, and, like that word, may be used as a verb to describe the activity.

OVERDEVELOP (PHOTOGRAPHY).

To allow a photographic **image** to **develop** either for too long a time, or by using too strong, or too warm, a chemical mixture. Whatever the cause, an overdeveloped picture will tend to exhibit too great a **contrast**.

OVER-DUBBING. See **sound-on-sound**.

OVEREXPOSE (PHOTOGRAPHY).

To allow a **film** to be exposed to light too long. The result is a rather "washed out" looking **print**.

OVERFLOW.

The generation of so great an abundance of **data** that it cannot be properly displayed or stored because it exceeds the display, or **storage**, device's capacity. An example would be using a hand-held calculator with an eight digit display and attempting to display a result of nine or more digits.

OVERHEAD PROJECTOR.

A very effective teaching device which uses a light source to shine through a **transparency**, often called a **view graph**, usually a 10x10-inch sheet of transparent acetate containing printed, written, or drawn material which is placed on the platform of the projector. The **image** is then projected onto a **screen** behind the machine. The lens and mirror arrangement in the overhead projector makes possible a bright projected image on the screen in semidarkened rooms or even in rooms that are fully lighted.

OVERHEAD PROJECTOR (cont'd)

Two basic types of overhead projectors are currently in use. The most popular one has its light source located beneath a stage or platform. Light shines through the stage (and the transparency) from below to the projector "head." The head contains a mirror, which changes the direction of the light beam and projects the image onto the screen. The second type, produced by the 3M Company, is probably best suited for use with small groups. The light source is in the top of the projector and shines down through the transparency to a substage mirror. It is then reflected back to the head from the mirror at a slightly different angle, and another mirror in the top reflects the image onto the screen. In operating either type of overhead projector, you must consider and compensate for the **keystone effect**.

Two important factors to keep in mind when purchasing an overhead projector are stage dimensions and **lumen** output. Although there are a great many 5x5-inch and 7x7-inch stage projectors in use, the best size for the classroom is the 10x10-inch. It is most effective for projecting pictures, charts, printing from books and articles, etc., which are suitable for direct transfer to 10x10-inch transparencies by relatively easy copy methods.

The power of some overhead projectors is expressed in watts, and much of the stated wattage may be dissipated in the form of heat. Thus, the lumen output of a lower wattage quartz-iodine **projection lamp** (which has less heat dissipation) may be greater than a higher wattage **incandescent light** projection bulb.

OVERLAP (PHOTOGRAPHY). *See* **multiple exposure.**

OVERLAY. *See* **transparency.**

OVERLOAD. *See* **over modulation.**

OVER MODULATION.

This usually means that the **amplitude** of the material being recorded (whether on **magnetic tape** or **phonograph disc**) is too great and causes **distortion** when played back. In **magnetic recording**, over modulation frequently causes **print through**. In television, over modulation may cause the phenomenon known as **halo**.

PBS. *See* **Public Broadcasting Service.**

PCMI. *See* **photo-chromic-micro-image.**

PM. *See* **maintenance.**

PROM. *See* **programmable read only memory.**

PTV. *See* **public television.**

PACKING DENSITY.

The number of **data** units a **medium** of **storage** can hold in relation to a specified dimension. An example would be the number of **bytes** per linear inch a specific **magnetic tape** may be able to record (this specific example is used often enough that the abbreviation bpi is commonly used for bytes-per-inch).

PAGE PRINTER.

A **printer** which prints a complete page at a time as compared to a **line printer** or a printer which prints a single **character** at a time.

PANCHROMATIC FILM (PAN).
Black-and-white **film** which is sensitive to all the colors in the visible portion of the spectrum and which renders those colors' relationships accurately in shades of black, gray, and white.

PAN HEAD. *See* **tripod (photography)**.

PANNING.
The act of moving a camera in an arc (usually parallel to the ground) in order either to **film** a panoramic view, or to "stop" the motion of a moving object (e.g., horse, car, etc.) by moving the camera in the same direction of motion to eliminate potential blurring of the moving object.

PAPER COPY. *See* **hard copy**.

PAPER TAPE.
An older **medium** of **storage**, now obsolescent, which used a long strip of paper in which holes were punched across its width in order to encode **data**.

PARALLAX.
In cameras using an **optical viewing system** this refers to the discrepancy between the **image** seen in the **viewing system** lens and the image transmitted to the **film** by the "taking" lens. The **single lens reflex camera** eliminates parallax.

PARALLEL TRANSMISSION.
The transmission of **data** or **information** in which the **character**s are transmitted simultaneously over several separate communications lines. Sometimes called simultaneous transmission. Contrasted with **serial transmission**.

PARITY.
An elementary method of error detection in which a "non-**data**" **bit** is added to the data bits forming a **character** so that the total number of "one" bits is always even or odd.

PASSIVE RADIATOR.
A **loudspeaker** which is mounted in an airtight enclosure with an "active" loudspeaker. Like the active loudspeaker, the passive radiator faces the listening area and radiates sound, but it is driven by pressure waves (sound waves) created within the enclosure by the active loudspeaker, not by electrical **signal**s from the **amplifier**.

PASSWORD.
A carefully specified, and usually confidential, set of **character**s (ordinarily eight or fewer) which a user may be required to **input** in order to use a **computer** system. The password is one means of security, that is, it maintains both the privacy and integrity of the **data** or **information** the system may offer.

PATCH CORD.
A **coaxial cable** used to interconnect two pieces of equipment. Usually a patch cord has **plug**s at both ends, but it may have a plug at one end and a **jack** at the other, or both ends may terminate in jacks.

PATCHING.
This means the connecting of two or more devices by means of **patch cord**s.

PAUSE CONTROL.

A control switch on a **tape recorder** which, when activated, stops the device in the record and/or playback **mode**. Usually, the pause control is a dual action switch—i.e., when it is pressed once, the machine stops almost instantly, and when it is pressed again, the machine starts almost instantly. This allows the user to eliminate unwanted portions of the **program** being recorded. For instance, a television broadcast being recorded may have commercials between segments; by proper use of the pause control, these need not be recorded. When used in this way, the pause control allows for a kind of simple **editing**.

In playback, one may wish to stop the tape for discussion, and then resume from the point at which the tape was stopped. Here too, the pause control is handy.

PAWL-SPROCKET.

A **film** transport mechanism in a camera or projector, which is located immediately below (or, in some cases, in) the **gate**. It allows for rotation in only one direction and advances the film one **frame** at a time by engaging with the **sprocket** holes at specific intervals. The pawl-sprocket (or claw) works together with the **shutter**. Just as it engages the film perforation, the shutter starts to close. When the light is cut off, the claw pulls down the film. Then the shutter uncovers the next frame as the claw disengages the film.

PEAK-HOLD. See **peak meter**.

PEAK METER.

A **signal strength meter** which indicates **amplitude** peaks more accurately than can a **VU meter**, this to allow more accurate recording of the loudest **signal**s. Peak meters are usually composed of **light-emitting-diodes** or **fluorescent light** bulbs. Peak-hold meters are those which, when so set (usually by depressing a switch), will continue to display peak readings until the "hold" setting is cancelled (usually by depressing the "set" switch).

PEAK POWER. See **IHF power**.

PERIPHERAL EQUIPMENT.

Any **hardware** in a **data processing** system which is not under the continuous control, or part of, the **central processing unit**. Examples are an **off-line printer**, a **terminal**, and a **modem**.

PERMUTED INDEX. See **keyword**.

PERSONAL COMPUTER. See **microcomputer**.

PHASE.

A term used to refer to waves pulsating in **synchronization** with each other. In **audio** it usually refers to either the **diaphragm**s of two or more **microphone**s or the **cone**s of two or more **loudspeakers**. When the cones, for example, are all moving forward and backward in unison, they are said to be in phase. If any one of these is not in synchronization, obviously they are then out of phase. The same is true for the diaphragms of the microphones. When such devices are out of phase, there is a resultant loss of the potential **frequency range**, particularly in the **bass**, plus, in **stereophonic** or **quadraphonic**, a confusion of **channel separation** and directionality.

In order to correct a pair of out-of-phase speakers, one simply reverses the connections to one of the speakers, not to both.

PHASE REVERSE CONTROL.

On **stereophonic** devices like **amplifiers** and **receivers**, this allows changing the **phase** of the **loudspeakers** in order to obtain proper phasing without physically reconnecting the cables to the loudspeakers.

PHONO CARTRIDGE.

A **transducer** that changes the wave forms in the grooves of a **phonograph disc** from mechanical energy into electrical impulses, which are fed to a **preamplifier** or **amplifier**.

There are two basic types of phono cartridge: the **magnetic cartridge** and the **piezoelectric phono cartridge**. The type most often used today is the magnetic cartridge. Piezoelectric phono cartridges are most often found in cheap equipment.

The phono cartridge uses a **stylus** to retrace the wave forms in the grooves of the phonograph disc. Occasionally, the phono cartridge may be referred to as the pickup.

PHONOGRAPH.

A device designed to reproduce the sound recorded on a phonograph **disc**. A phonograph is usually a single unit containing a number of components and it may be able to reproduce the sound from **monaural, stereophonic,** or **quadraphonic** discs.

Phonographs are frequently found as portable, self-contained units, but every such phonograph must have the following components: **turntable, tone arm, phono cartridge, amplifier**(s), and **loudspeaker**(s).

A phonograph that is designed to be portable should have a tone arm that is lockable, so that it does not flop around while the machine is being transported.

PHONOGRAPH DISC.

A flat, round, plate-like object on whose surface is incised a spiral groove in which are undulations which are **analogs** of sound. In the center of the plate is a hole which fits on the spindle of a **phonograph**. The most common playing speeds currently in use are 33⅓ and 45 revolutions per minute (rpm). Older consumer discs were of 78 rpm, while broadcast (or "transcription") discs were made to play at 16 rpm.

The usual diameters for regular phonograph discs are either twelve inches or ten inches (for both 33⅓ rpm and 78 rpm) and seven inches for 45 rpm.

PHONO INPUT.

A **jack** on an **amplifier** or **preamplifier** which accepts the **plug** from a **phono cartridge**. Some devices have two such **input** jacks, one for **magnetic cartridges** and the other for **piezoelectric phono cartridges**. The former will be marked as "mag" while the latter may be marked as "Xtal" or "crystal." A mismatch in **impedance** and voltage will occur if a piezoelectric phono cartridge is plugged into the circuit of a magnetic cartridge phono input (which is what most phono inputs on newer equipment are).

PHOTOCELL. *See* **photoelectric cell**.

PHOTO-CHROMIC-MICRO-IMAGE (PCMI).

A **microform** development of The National Cash Register Company, this is a method of photo-reduction that allows a huge number (approximately three thousand) of pages to be stored on a 4x6-inch transparent sheet. PCMI offers excellent resolution of the stored **image**s and is comparatively hardy in its resistance to scratching, to attack by bacteria, or to problems caused by high humidity or temperature. This system is frequently called ultrafiche.

PHOTOCOPY.
A copy of a document made by a modern **copying machine** but so called because, before the advent of such devices, the only copying method available was photographic.

PHOTOELECTRIC CELL.
A light-sensitive device used for converting variations of light intensity into electrical **signal**s. The photoelectric cell is found in **motion picture projectors** that take **film** with an **optical sound track.** ˙

Also a cell which converts the energy of light into electric energy. It is often a part of an **exposure meter**.

PHOTOEMISSIVE.
Emitting or capable of emitting electrons upon exposure to radiation in and near the visible region of the spectrum, i.e., light.

PHOTOGRAPHIC PAPER.
Paper which has a special surface, or **emulsion**, which is light sensitive and which can be used to make photographic **print**s.

PHOTOMACROGRAPH.
A photograph whose subject is represented as larger than the subject is in reality. A photomacrograph can be made either by photographing the subject at close range to the lens (a special **close-up,** or "macro" lens is often used for this purpose) or by increasing the size of the **image** through the use of an **enlarger**.

PHOTOMETER. *See* **exposure meter**.

PHOTOMICROGRAPHY.
The technique of taking photographs through a microscope, which produces ultra **close-up,** enlarged views of microscopic life or specimens. With special equipment and color **film**, **slides** can be produced that can be shown by an ordinary 2x2-inch **slide projector**. A wide range of such slides is available commercially.

PICKUP. *See* **phono cartridge**.

PICKUP TUBE.
An electron-beam tube used in a television camera wherein an electron current or a charge-density **image** is formed from an optical image and scanned in a predetermined sequence to provide an electrical **signal**. The two major types are the **orthicon image camera tube** and the **vidicon camera tube**.

PICTURE-GRADE CHANNEL. *See* **video-grade channel**.

PICTURE MONITOR (TELEVISION). *See* **monitor**.

PICTURE TRANSFER (LIFT PROCESS). *See* **transparency**.

PICTURE TUBE, TELEVISION. *See* **cathode-ray tube**.

PIEZOELECTRIC PHONO CARTRIDGE.

A type of **phono cartridge** that is opposed to the **magnetic cartridge**. It uses a material that, when deformed slightly by physical pressure, emits an electrical voltage proportional to the stress.

There are two types of piezoelectric phono cartridges: the crystal (sometimes "Xtal") kind, usually made with crystals of Rochelle salt; and the more recently developed ceramic type. The crystal cartridge generally has poor **compliance** and poor **frequency response** and is ordinarily found now only in cheap **phonograph**s made for children. Ceramic cartridges are better than the crystal ones and are not infrequently used in less expensive phonographs.

Besides having poorer compliance and frequency response than good magnetic cartridges, piezoelectric phono cartridges can be easily destroyed (they are almost never designed to be repaired) by rough handling, extremes in temperature and humidity, and wear.

As a general rule, devices that use or are suited to the magnetic cartridge are preferable to those that require a piezoelectric phono cartridge.

PINCH ROLLER. *See* **capstan**.

PISTON SPEAKER.

The most commonly encountered type of **loudspeaker**. The piston speaker (sometimes called dynamic speaker) is a **transducer** that is rather simple in its design. The electrical **signal** from the **amplifier** is fed to a coil of wire (the "voice coil") which is suspended between the poles of a large magnet. Attached to the voice coil is a large paper or plastic **cone** (or piston), whose outer edges are attached to a rim of metal. As the electrical impulses from the amplifier enter the voice coil, they cause the coil and, in turn, the cone, to vibrate—thus passing these vibrations on to the surrounding air and creating sound waves. The most usual alternative to the piston speaker is the **electrostatic speaker**.

PLAIN PAPER COPYING. *See* **electrostatic copying**.

PLANETARY CAMERA.

A camera used in making **microform**s. The document to be copied and the **film** in the camera remain stationary during the process of **exposure**. The planetary camera is usually contrasted with the **rotary camera** and the **step-and-repeat camera**.

PLATEN. *See* **opaque projector**.

PLATTER. *See* **turntable platter**.

PLAYBACK DECK.

A **deck** that plays **magnetic tape** and that can be used only to play back the tape, and not to record on it. This is intended to save the cost of various components in the deck, and to prevent the unintentional **erase** of a pre-recorded tape.

PLAYBACK HEAD.

A **head** in a **tape recorder** that converts the recorded magnetic **signal**s on the tape into electrical impulses and sends them to the playback **amplifier** and from there to the **loudspeaker**, etc.

In an **audio** device with three heads, the playback head is third, following the **erase head** and **record head**.

PLAYBACK HEAD (cont'd)

In some cheaper machines—and in most audio **cassette** devices—the record head and the playback head are one and the same. Such machines cannot offer the ability to **monitor** off the tape.

Further, some devices intended only to play back **magnetic tape**, and not to record upon it (e.g., a **playback deck**), will use only a playback head. There are many **videotape** and audio cassette devices so designed both for cost savings and to prevent tapes from being inadvertently **erased**.

PLOTTER.

A piece of **peripheral equipment** which is capable of drawing (plotting) graphs of charts automatically as **data** or **information** is fed to it.

PLUG.

A connector that receives the **jack**. The most common are the one-quarter-inch phone plug, the phono (or, commonly, RCA) plug, the miniature plug, and the subminiature plug. For institutional use, the large (and strong) one-quarter-inch plug is recommended.

POLARITY (PHOTOGRAPHY).

The relationship of tones in a photographic **image** which, when they are inverse to the original subject's tones are called **negative** and when not inverted are called **positive**.

POLARITY (TELEVISION).

This refers to the polarity of the black portion of the picture **signal** with respect to the white portion of the picture signal. For example, in a "black **negative**" picture, the potential corresponding to the black areas of the picture is negative with respect to the potential corresponding to the white areas of the picture, while in a "black **positive**" picture the potential corresponding to the black areas of the picture is positive. The signal as observed at broadcasters' master control rooms and telephone company television operating centers is "black negative."

POLARIZED PROJECTION.

A technique whereby the illusion of motion can be created on transparencies through the use of light-polarizing materials (technamation materials) which are applied directly to the surface of the **transparency film**. When placed on the **overhead projector**, the **image** from the transparency passes through a rotating "spinner" of polarized glass. When the image is seen on the **screen**, wheels seem to be turning, fluids flowing, etc.

POLARIZING FILTER. *See* **filter (photography)**.

POLAR PATTERN.

A pattern drawn through the horizontal or vertical plane which is used to depict graphically either the **dispersion** of a **loudspeaker** or the area of space which a **microphone** is sensitive to (or "hears").

One type of polar pattern designation for a microphone is the **cardioid pattern**.

POLLING.

In a **time share computer** system, polling is a method of querying each **channel** in order for the system to receive incoming **data** in a logical and fair fashion.

POLYESTER TAPE.

A **base** for **magnetic tape** designed for **magnetic recording**. Its makeup is a strong, transparent plastic like mylar. Unlike **acetate tape**, which becomes brittle with age, polyester tape retains its flexibility indefinitely. The only disadvantage of polyester tape is its elasticity—it can be easily stretched unless it is **pretensilized tape** (i.e., prestretched by the manufacturer). Polyester tape should be stored in a relatively dry atmosphere whose temperature range is from fifty to eighty-five degrees Fahrenheit. For storage, polyester tape should not be too tightly wound, and it should be stored vertically.

POP SCREEN. See **microphone**.

PORT.

Generally, the entrance to, or exit from, a computer **network**. Specifically, a port is usually intended to mean the entrance and exit for a **data** channel in a **computer**. Each port is usually dedicated to one **channel**.

PORTED BAFFLE. See **baffle**.

POSITIVE.

A **film**, or **print**, in which the **image** accurately renders the tonal values of the original subject. When these tonal values are reversed, the film, or print, is said to be a **negative**.

POUNCE PATTERNS.

The results of a technique whereby the outline of a map, silhouette, or other shape is drawn on a stencil (usually paper). Holes are then poked through the outline at appropriate intervals. The stencil is placed on a clean **chalkboard** and a chalk eraser is dusted over each of the holes. When the stencil is removed, the dotted outline of the figure can be connected by chalking lines from one point to another.

POWER AMPLIFIER. See **amplifier**.

POWER BANDWIDTH.

A specification for an **amplifier** which expresses the **frequency range** of the device at its rated **power output**. A power bandwidth specification for an amplifier could be twenty **hertz** to twenty thousand hertz at twenty watts per **channel**, with 1 percent **harmonic distortion**. This last portion is imperative for a full expression of the specification.

POWER CORD. See **line cord**.

POWER OUTPUT.

A statement, ordinarily in watts, of the amount of electrical **signal** power an **amplifier** can produce. Power output can be expressed in various ways, but the most reliable expression is as **continuous power** (or RMS power, for root-mean-square power). Such terms as **IHF power** or music power are misleading.

PREAMPLIFIER.

A separate device, or a circuit within an **amplifier**, which takes a **signal** of relatively small **amplitude** and amplifies it sufficiently so that it can be used by a power amplifier. The preamplifier deals with such signals as those emanating from **magnetic cartridge**s, **tuner**s, and **microphone**s.

PRE-ECHO.

A phenomenon that occurs in **phonograph discs** and occasionally in **magnetic tape** recordings. The term stems from the fact that, in playing quiet passages of such recordings, one hears the **program** about to be played back both faintly and slightly before the actual program occurs. This happens in phonograph discs either because the walls of the grooves of the disc are too thin (usually due to **over modulation**) or because the magnetic tape master from which the disc was made had been recorded with too much **amplitude**, or the tape **base** was too thin to prevent **print through**.

PRESSURE PAD.

A device on a **tape recorder** that physically presses the **magnetic tape** against the **head** assembly (to ensure good tape-to-head contact) in the record and playback **modes**. Good tape recorders often do not use pressure pads, since they can cause rapid (and excessive) head and tape wear. Pressure pads should be automatically disengaged in the **rewind** and **fast-forward** modes to minimize wear.

PRESSURE ROLLER. *See* **capstan**.

PRETENSILIZED TAPE.

A **polyester tape**, used in **magnetic recording**, which has been prestretched by the manufacturer in order to minimize further stretching.

PREVENTIVE MAINTENANCE. *See* **maintenance**.

PRINT (PHOTOGRAPHY).

The act, or product of that act, of creating an inverse copy (usually a **positive** from a **negative**) of a **film**, the copy being made on a special photosensitive paper ("print" or "printing" paper). The two usual ways of creating a print are through the use of an **enlarger** or by making a **contact print**.

PRINTER.

A piece of **peripheral equipment** whose function it is to accept **digital data** and create an eye-legible printed record of that data (i.e., a **printout**). There are two basic types of printer: the **thermal printer** and the **impact printer**, the latter tending to be the faster (and noisier). A new type of printer in use with **main frame** computers is one which uses **electrostatic copying** technology (and is termed, not surprisingly, an electrostatic printer). It is able to print at rates up to thirty thousand lines-per-minute (or five hundred lines, or eight pages, per second).

PRINTOUT.

The product of a **printer**: the eye-legible printed record of **data** converted from **digital** form into recognizable **characters**.

PRINT THROUGH.

This is usually evidenced as an echo (or **pre-echo**) effect in **magnetic tape**. It is caused when a loud passage on one layer of tape on a **reel** impresses itself (by magnetization) on an adjacent layer, either above or below it. It can be minimized by using proper **amplitude** (i.e., volume) **level**s during recording and by using tape with a relatively thick **base** (e.g., 1-**mil** or 1.5-mils).

PROCESSED FILM.

Film which has been so treated as to convert the **latent image** thereon to a visible **image**. To **develop** film is one method of processing it.

PROGRAM.

Either the process, or the result of that process, which prepares a set of instructions, statements or commands, which are in a form comprehensible to a **computer** and which are designed in order to achieve a specific set of results. In a sense, the program is a routine which the computer follows in order to provide the user with desired **data** or **information**. The program is usually written in a special **language**.

In broadcasting or recording parlance, a program is (as a noun) a self-contained unit to be broadcast, or (as a verb) the preparation, or actual broadcast, of such a unit.

PROGRAMMABLE READ ONLY MEMORY (PROM).

A small **memory chip** which is not imprinted with a **program** at the time of its manufacture, but rather by the user. This type of memory can usually only be programmed once and can store, but not act upon, the **data** it contains. Contrasted with the **erasable programmable read only memory (EPROM)**, the **read only memory (ROM)**, and the **random access memory (RAM)**.

PROGRAMMED INSTRUCTION (P.I.).

An educational method or technique designed to implement individualized or tutorial instruction. It has also been described, by Lawrence M. Stolurow, as a philosophic approach to the education of the individual student, which is based upon a psychological analysis of the teaching-learning process. P.I. consists of "a planned sequency of experiences, leading to proficiency, in terms of stimulus-response relationships." This is the **program**, and it has usually been tested to prove its effectiveness.

In a program the subject matter is broken up into small units called **frames**. At least part of the frame requires some type of response from the student. Upon reacting, the student is provided with immediate **feedback** — i.e., he is told the correctness of his answer. Programs are aimed at specific goals and the units are arranged in careful sequence to achieve the desired objective. Thus, programmed instruction is a process that especially lends itself to a **systems analysis** or a **systems approach**.

Two basic types of programs can be used, depending on the kind of response demanded from the student. The linear or constructed-response program requires the student to write an answer to each question in the program. In this type, the sequence of frames seen by one student is seen by all other students and they all proceed straight through the program.

A second general form, known as intrinsic programming, has questions that serve a diagnostic purpose. Its technique adapts to the individual student and allows him to branch off from the main sequence of frames either for remediation or for acceleration. The adaptive or branching program is one in which the sequence of frames seen by each student is determined by his responses to questions asked by the program — usually multiple choice questions. Thus, certain sequences may be by-passed and the program allows for greater differences in student abilities.

A combination program can be devised which uses both the linear and adaptive types.

Most programmed materials are presented in book form **(software)**, but some are designed for the "teaching machine" **(hardware)** — a device for presenting programmed instruction. Much of the significant early work in the development of teaching machines and programs was done by Sidney L. Pressey at Ohio State University and B. F. Skinner at Harvard. Pressey came out with his first teaching machine in 1926 and Skinner developed a number of devices for constructed-response programs.

PROGRAMMED INSTRUCTION (P.I.) (cont'd)

Computer assisted instruction is perhaps the ultimate method for implementing programmed instruction. Special programs can be set up in a computer, which then responds to **input**s made by a student on an electric typewriter or other means. One of the first attempts at combining the two was made by IBM in 1958. This teaching machine combined an IBM 650 digital computer and an electric typewriter which input and programmed the computer to teach binary arithmetic. In 1961, a Bendix G-15 computer was adapted to a complex multiple choice machine which used a specially built random-access **slide projector** and/or an electric typewriter input. The computer was used to control the branching. At present computers may be used to teach both linear and adaptive programs.

PROGRAMMING LANGUAGE.

The **language** used to write a **program** for a **computer**.

PROJECTION LAMP.

This is used in various types of projectors to illuminate the **film** or **slide image** and to project it upon the **screen** via the lens. Formerly, most 16mm **motion picture projectors** used an **incandescent light** tungsten lamp, which could be run from a normal **alternating current** power supply. These were usually 500 watt or 750 watt lamps, and they created a great deal of heat. Now, tungsten-halogen lamps are used for both the 16mm and 8mm projectors because they provide an improved source of light and are more efficient in their use of electrical power.

PROJECTION LENS. *See* **lens, projection**.

PROJECTION RATE. *See* **frames-per-second**.

PROMPT.

An instructional message given a **computer** user usually in order to assist that user in making decisions as to offered options. The prompt is generated by the **program** being used and is provided automatically.

PROOF. *See* **contact print**.

PROTOCOL.

A set of formal conventions or procedures which determine or govern the format or timing of message exchanges between two or more communicating (and usually different) **computer** systems. Among protocol procedures are log-on, **password**, and log-off.

PUBLIC BROADCASTING SERVICE (PBS).

An agency of the **Corporation for Public Broadcasting** which manages the **public television** network by coordinating the operations of public television stations and developing national **program**ming.

PUBLIC TELEVISION (PTV).

A general term referring to a type of television **program**ming of human interest and importance which is usually not available on commercial television. However, the creation of the **Corporation for Public Broadcasting** in 1967 for the purpose of encouraging the growth and development of noncommercial radio and television expanded the scope of public television to include the use of such media for instructional purposes. Thus, public

136 / PUBLIC TELEVISION (PTV)

PUBLIC TELEVISION (PTV) (cont'd)
television in this sense refers to the operations of the Corporation for Public Broadcasting which, through the **Public Broadcasting Service** and other agencies, manages and assists a number of regional and state educational television **network**s.

PUCK. *See* **idler wheel**.

PULSE DIAL. *See* **rotary dial**.

PUNCHED CARD.
A card, like the **Hollerith card**, which carries **digital** type **data** by means of holes punched into the body of the card. Punched cards are encoded by a **card punch**, can be read by a **card reader**, and may be arranged by a **card sorter**.

QS. *See* **matrix four channel**.

QUADRADISC. *See* **CD-4**.

QUADRAPHONIC.
Although now obsolescent, in **audio** this describes a system (or recording) which provides four **channel**s of sound, separately and simultaneously. **Stereophonic** sound systems have just two such channels: left and right. In the quadraphonic system, the four channels are designated as front right and front left, rear right and rear left. There are three basic types of quadraphonic systems: the **CD-4**, **discrete four channel** tape, and **matrix four channel**.
The advantage claimed for quadraphonic listening, as opposed to either **monaural** or stereophonic listening, is that it is better able to recreate the **ambiance** of the actual environment where the **program** was made, and is able to include such subtleties as natural reverberation, total "sound surround."

QUARTER TRACK.
A method of **magnetic recording** either **monaural**, **stereophonic**, or **quadraphonic program**s on **open-reel** type **magnetic tape**, or stereophonic **program**s on audio **cassette** tape. In all, four **track**s are recorded, each track being slightly less than one quarter of the width of the tape.
Open-reel quarter track monaural recordings contain four tracks of separate **information**. In order to be played back, they must be reversed after each track has been played, since two tracks (1 and 3) run in one direction and the other two (2 and 4) run in the other. Stereophonic open-reel quarter track tapes contain two pairs of interlocking (1 and 3, 2 and 4) stereophonic tracks. After one program has been played back, the tape must be reversed to play the other. A quadraphonic quarter track open-reel tape contains four tracks running in one direction, all four constituting one program.
An audio cassette stereophonic tape contains two pairs of stereophonic tracks (tracks 1 and 2, and tracks 3 and 4); one pair runs in one direction, while the other pair runs in the opposite direction, thus after one pair has been played, the tape must be reversed in order to play the other.

QUEUE.
A group of people, messages, transactions, etc., arranged conceptually in a line, with the most recent arrivals at the rear of that line. In a sense, a queue is a waiting list.

RAM. *See* **random access memory**.

RCA FOUR CHANNEL. *See* **CD-4**.

RCA PLUG. *See* **plug**.

RIAA. *See* **Recording Industry Association of America**.

RIE. *See* **Educational Resources Information Centers**.

RJE. *See* **remote job entry**.

RMS POWER. *See* **continuous power**.

ROM. *See* **read only memory**.

RADIAL ARM.
A **phonograph**'s **tone arm** whose design is such that, rather than describing an arc across the surface of the **phonograph disc**, it follows a straight line path through the radius of the disc. Although the radial arm **turntable** is rather more expensive than the conventional tone arm turntable, the **stylus** in the radial arm is always at right angles to the grooves of the **disc**, and there is essentially no **tracking error** and the need for **anti-skating** compensation is eliminated.

RADIO, AM. *See* **amplitude modulation**.

RADIO, FM. *See* **frequency modulation**.

RANDOM ACCESS.
The ability to **access**, or locate, specific **data** in **memory** or **storage** directly, that is without having to pass through data which may be located on either side, so to speak, of the desired data. In a sense, a **phonograph disc** provides random access since the **stylus** for playback can be placed anywhere on the surface of the **disc** and, therefore, the entire disc need not be played in order to hear only the last portion (i.e., that section closest to the label). Random access may be contrasted to **serial access**. Serial access requires passing through all intervening data in order to locate the required section. The analogy here is much like a **magnetic tape** which has to be gotten through (even though getting through may be done in the **fast forward** or **rewind modes**) to arrive at the point required.
Those types of memory or storage which provide random access are **core**, **chip**, **disk**, etc., as opposed to magnetic tape.

RANDOM ACCESS MEMORY (RAM).
Memory which is of a **random access** type, usually on **chip**s, and which, in **computer** terminology, can both **read** and **write**. Ordinarily, random access memory of this type can be reprogrammed with new **program**s or **data** an indefinite number of times.

RANGE FINDER CAMERA.
Any camera whose **viewing system** incorporates some means for semi-automatically determining the distance from the camera to the subject. Ordinarily, the out-of-focus **image** is seen in such a viewing system as either a double image (i.e., one image of the subject superimposed upon another) or as fuzzy. Some range finder cameras are of the

RANGE FINDER CAMERA (cont'd)

"split image" type. Here, a vertical line appears in two sections when the line is out of focus. When properly focused, depending on the type of range finder, the image will appear as a single one, or it will no longer be fuzzy, or it will appear as a continuous line.

The range finder camera is very often a **reflex camera**.

RASTER.

The screen of a **cathode-ray tube** used in television sets, etc., which is scanned by the electron beams in order to produce the picture.

RATED POWER OUTPUT.

The maximum power (usually given in watts) of an **amplifier** as stated by the manufacturer. A more meaningful specification is either **continuous power** or the **power bandwidth**.

RAW STOCK (PHOTOGRAPHY).

Photosensitive material (i.e., **film, print** paper, **electrostatic copying** paper, etc.) which has not been subjected to **exposure**.

READ (COMPUTER).

The ability of a device to either interrogate the material in **memory** (without effecting any change in those contents) or to transfer **data** from one form of **storage** to another.

READER.

A device which incorporates a function of an **enlarger**, in that it projects an enlarged **image** from a **microform** onto a **screen**, in order that the microform can be eye-legible. Some readers are designed solely for **microfilm**, some only for **microfiche** and some only for **microcards**. More often, however, readers are designed to accept either microfilm or microfiche.

Of the readers specific to microfilm use, there are essentially two types: manual (or "hand cranked") and motorized.

Many readers are now so constructed that they may fold into a self-contained carrying case not much larger (but somewhat heavier) than a brief case. Naturally, these are called portable readers.

READER/PRINTER.

A **reader** for **microform**s which can make paper copies which closely approximate the size of the projected **image**. The variety of reader/printers is almost as great as that of readers, and the **copying machine**s incorporated into them are most often of the **electrostatic copying** type. A reader/printer is the same as an enlarger/printer.

READ ONLY MEMORY (ROM).

A **memory chip** which is imprinted with its **data** at the time of manufacture and can never be altered afterward.

READOUT.

The display of **data** or **information** on a **cathode-ray tube**. Compare to **printout**.

REAL TIME.

The processing and provision of **data** or **information** by a **computer** sufficiently rapidly that the results may be used to direct, control, or be otherwise integrated into actual

REAL TIME (cont'd)

ongoing processes. For example, the guidance of an aircraft's flight requires data or information in real time.

REALIA.

A fancy term for "real things" that can be used in classrooms, libraries, and museums. They may be unmodified real things, live or inanimate, which have been removed from their original real-life surroundings; or they may be modified real things, which have been changed in some way for educational purposes (e.g., a cut-away version of an automobile engine). Finally, specimens may be used (such as frogs in a biology class).

REAR SCREEN PROJECTION.

A technique of projecting **images** from **motion picture projectors**, a **slide projector**, or a **filmstrip** projector or viewer onto a piece of ground glass or a translucent **screen** from the rear. This permits an unobstructed view of the picture in front of the screen. Rear screen projection is frequently used in the wet **carrel**s of **media center**s.

RECEIVER.

An **audio** or **video** device that contains a **tuner** (to select either radio or television **channel**s), a **preamplifier**, an **amplifier**, **antenna**, various types of controls and, in the case of a television receiver, a **loudspeaker**.

RECORD CHANGER. See **turntable**.

RECORD HEAD.

The **head** in a **tape recorder** which converts the electrical **signal**s (from the **microphone**, etc.) into magnetic signals and records these onto the **magnetic tape**.

In an **audio** device with three heads, the record head is second, preceded by the **erase head** and followed by the **playback head**.

In some cheaper machines—and most audio **cassette** devices—the record head and the playback head are one and the same. With such machines, therefore, one cannot **monitor** from the tape itself.

RECORDING AMPLIFIER.

An **amplifier** in a **tape recorder** which increases the **amplitude** of the **input** signal sufficiently so that the **record head** may properly magnetize the **magnetic tape**, thus recording the **signal**.

Some tape devices do not include a recording amplifier (or record head) and can only be used for playback; these are usually termed **playback deck**s.

RECORDING DECK. See **deck**.

RECORDING INDUSTRY ASSOCIATION OF AMERICA (RIAA).

This is the organization which specified (in 1955) the **equalization** to be used by the **phonograph disc** manufacturers of the United States.

RECORDING TAPE TYPE. See **tape type**.

RECORD INTERLOCK.

A switch on a **tape recorder** which prevents the operator from accidentally putting the device into the record **mode**. Usually, the interlock has to be activated first, followed by the second (record) switch, while actually puts the device into the record mode.

140 / REDUCTION (PHOTOGRAPHY)

REDUCTION (PHOTOGRAPHY). *See* **diameter**.

REDUCTION RATIO. *See* **enlargement ratio**.

REEL.
A spool upon which **magnetic tape** or motion picture **film** is stored. In **open-reel** tape, the diameter of the reel may be 3, 4, 5, 7 (the most common size), or 10½ inches (studio size). The "reel" in an audio **cassette** tape is little more than a simple hub. Tape or film is fed from the **supply reel** to the **take-up reel**.
A standard film projection reel holds four hundred feet of 16mm sound film and has a running time of eleven minutes (at twenty-four **frames-per-second**).

REEL-TO-REEL. *See* **open-reel**.

REFLECTANCE.
The ability of a surface to reflect light falling on it. The greater the reflectance, the shinier the surface. Conversely, the less the reflectance, the duller the surface. In photographic **print**s, those with high reflectance are called **glossy**, those with low reflectance are called **matte**.

REFLEX CAMERA.
A camera in which the operator may see the area of the scene to be photographed through a view finder (*see* **viewing system**) which is coupled to the lens that actually takes the picture. In the **single lens reflex camera**, the view finder lens and the "taking" lens are one and the same, thus preventing the problem of **parallax**. In the **optical viewing system** camera (e.g., the twin lens reflex camera), the view finder lens and the taking lens are separate. As the operator focuses the view finder lens, a mechanical coupling focuses the taking lens. However, the optical viewing system camera, whether a reflex camera or not, is subject to parallax.

REFRACTIVE INDEX (LIGHT). *See* **index of refraction**.

REFRESH.
The act of restoring **data** or **information** which otherwise would fade from whatever **medium** in which it may be. For example, the display on a **cathode-ray tube** needs to be refreshed by the action of the electron beam, otherwise it would disappear. In some "dynamic" **memory** cells (i.e., **random access memory**) there is also the need to refresh stored data in order for the cells to retain their contents.

REGENERATE. *See* **refresh**.

REMOTE JOB ENTRY (RJE).
The **input** of **data** to the **computer** from some remote location via some **input/output** device.

REPROGRAPHY.
The science or art of the reproduction of documentary material whether by **copying machine** to make **hard copy**, or by microreproduction to create **microform**s, etc.

RESEARCH IN EDUCATION. *See* **Educational Resources Information Centers**.

RESOLUTION. *See* **resolving power**.

RESOLVING POWER.
The degree to which a photographic or television camera (or system) can discriminate and reproduce fineness of detail. It is analagous to the **acutance** of a **film**.

Unfortunately, testing for the resolving power of a piece of **hardware** is always limited by the resolving power of the film or **magnetic tape** used in the test. Film varies greatly in its ability to resolve detail, depending on whether it is "fine grain," "very grainy," etc. Magnetic tape varies in its ability to resolve detail based upon its width, **coating**, speed, etc.

In photography, a resolving power of 30 lines/mm (which is the way resolving power is specified) is about average. Good sharpness results with resolving powers above 50 lines/mm.

In **videotape** and television, resolving power is specified as lines-per-inch (and usually specified as taken at the center of the **raster**). About 200 lines-per-inch, or more, is sufficiently good resolving power for such equipment.

RESONANCE FREQUENCY.
The **frequency** at which something, usually a mechanical or electrical device, vibrates with maximum **amplitude**. Resonance frequency is an important specification since it indicates where, in the **frequency range**, a device exhibits this undesirable reaction. The resonance frequency, to be relatively harmless, should be of low amplitude and at such a high (or low) frequency that it is either not audible or is tolerable. For a **loudspeaker**, a desirable resonance frequency would be forty-five **hertz** or lower; for a **tone arm**, below ten hertz.

RESPONSE TIME.
The total elapsed time between entering an inquiry via a **terminal** and the receipt of the answer as provided by the **computer**. Response time includes communications time, processing time, etc., and is similar to **access time**.

RETRIEVAL.
The act, or product of that act, of finding **data** in **storage** and providing it, usually by means of a **terminal**, to a user.

REWIND.
Mode in a **tape recorder** in which the **magnetic tape** is quickly returned to the **supply reel**. During rewind the tape should not be in contact with the **head**s since, at fast speed, the tape can induce excessive head wear. The opposite of rewind is **fast forward**.

RIGHT JUSTIFY. *See* **justification**.

RIM DRIVE.
A system of conveying rotational motion from the motor to the **turntable platter** in a **phonograph**. Interposed between the inside rim of the platter and the motor is an **idler wheel** which, when rotated by the motor, imparts that movement to the rim.

For various reasons, particularly a loss of speed accuracy, the rim drive is considered a less desirable system than either the **direct drive** or the **belt drive** system.

ROLL. *See* **roll-over**.

ROLL FEED. *See* **web feed**.

ROLL FILM.

Film of some length, usually with a **frame** size of 35mm or 16mm, which is stored on a **reel** (roll). This is the usual format for **microfilm** or motion picture film, as opposed to **microfiche** or **microcard**s.

ROLL-OVER.

An upward or downward movement of a television picture caused by a lack of vertical **synchronization**. If the synchronization loss is within the tolerances of the vertical synchronization circuit, roll-over can be controlled by using the "vertical control" of the device. Roll-over is sometimes called flip-flop or flop-over.

ROOT-MEAN-SQUARE POWER. See **continuous power**.

ROTARY CAMERA.

A **microfilm** camera which, as distinguished from a **planetary camera** or a **step and repeat camera**, photographs the documents as they are automatically conveyed past the lens. The document conveyer and the **film** in the camera move in **synchronization** so that no blurring of the **image**s occurs.

ROTARY DIAL.

Also known as pulse dial, rotary dial is the method of sending a **signal** in a telephone system in order to place a call. The dial is rotated from the appropriate hole to the stop and, as it returns to its rest position, generates a specific number of electrical impulses. Each set of impulses constitutes one **character**. Rotary dial need not employ a mechanical rotating dial since, recently, push button phones have been produced which generate the appropriate impulses when a button is depressed. Contrasted with **tone generated dial**.

RUMBLE.

An undesirable, low **frequency** sound, usually regular, which occurs in **phonograph** systems. It is caused either by vibrations in the **turntable platter** or motor of the **turntable**, or by some eccentricity of the phonograph **disc** itself (e.g., warpage, or a faulty manufacturing process which actually recorded the rumble sound into the grooves of the disc).

Rumble is also a specification for a turntable, indicating the amount of such unwanted noise in **decibel**s. It may be given either as "weighted" or "unweighted." The weighted rumble specification takes into account the fact that human hearing is not as acute in the lower **audible frequency range** as in the higher portions; therefore, the weighted figure tends to be a better one. The higher the number, the less the rumble. For a good turntable, the unweighted rumble should be at least forty-five **decibel**s; the weighted rumble should be a minimum of fifty-five decibels.

RUMBLE FILTER. See **filter switch**.

S/N RATIO. See **signal-to-noise ratio**.

SQ. See **matrix four channel**.

STV. See **school television**.

SAFELIGHT.
A light by means of which photosensitive materials (e.g., **film**) may be illuminated without undergoing **exposure** because such materials are not sensitive to the color or intensity of the safelight.

SCANNER. *See* **optical scanner**.

SCANNING LINE.
In television, a single, continuous strip which includes shadows, **half tone**s, and highlights which, taken with all the other scanning lines, make up the total picture. In American broadcast television, there are 525 interlaced scanning lines per **frame** (at thirty **frames-per-second**). The number of scanning lines is fixed; thus, as the size of a television set's **cathode-ray tube** (picture tube) is increased, the lines become more apparent because they are spaced more widely apart.

For **closed circuit television** and **videotape**, etc., the number of scanning lines may vary. However, it is usually conceded that systems that use fewer than 200 scanning lines fail to provide sufficient **resolving power**.

SCANNING RADIO.
A radio capable of scanning (i.e., checking from station to station automatically to determine which ones are carrying broadcast **information** or **signal**s) a given **frequency band** (e.g., citizens band) or set of frequency bands (e.g., police, fire, weather, aircraft, **frequency modulation** (FM), etc.). When a signal is encountered, the scanning stops so that the listener may hear the broadcast.

SCANNING SPOT (TELEVISION).
This refers to the cross-section of an electron beam at the point of incidence in a television camera tube or picture tube **(cathode-ray tube)**.

SCANNING (TELEVISION).
The process of breaking down an **image** into a series of elements or groups of elements representing light values and transmitting this **information** in time sequence.

SCHOOL TELEVISION.
Television **program**ming that pertains to the operation and purposes of school systems. This would include the use of television for direct instruction, school administrative purposes, teacher in-service training, and community **information** and education.

SCRATCH FILTER. *See* **filter switch**.

SCREEN.
Two basic types of screens are used for front projection: wall-mounted screens are pulled down for use and retracted into a case when not needed; portable stand screens are mounted on **tripod**s or detachable legs. These can be located where desired.

The type of surface a screen has is very important in terms of **resolution**, brightness, and angle of view.

The **matte** white screen gives a wide viewing angle and is especially useful when the audience must be seated close to the screen. It gives maximum resolution at short distances. Since it is used for close viewing, its relatively poor reflective ability or brightness is not critical.

SCREEN (cont'd)

The beaded screen has a much greater reflective power than the matte screen, but the brightness is mainly concentrated along the projection axis and fades to an inadequate level beyond forty degrees from the axis.

A smooth silver screen is well-suited to color reproduction because its metalic particles emphasize colors and increase brightness above the level of the matte white screen, but viewing angle and brightness are slightly less than modern types of beaded screens.

Lenticular screens have perhaps the best screen surface developed to date. Lenticulation (lens-shaped reflective particles) may be embossed in a vertical or horizontal direction, or in both. They combine good color quality, high reflectivity, and a fairly wide viewing angle from forty to sixty degrees. Also, there is virtually no loss of light at any viewing angle within the range.

There are also devices for **rear screen projection**.

SEARCH KEY.

A **character**, or group of characters, which, when **input** via an **input/output** device, will identify, locate and **output** desired **data** or information. An example of a search key may be "SAU =" which would then be followed by a personal name. This search key could mean provide me with all bibliographic entries authored by the personal name following the search key.

SEARCH (RADIO). *See* **scanning radio**.

SECONDARY STORAGE. *See* **auxiliary storage**.

SELECTIVITY.

A specification applied to equipment like **tuners** and **receivers**. The term means that the device can tune in a broadcast station without picking up **interference** from stations close to the desired one on the radio or television **band**. Selectivity is specified in **decibels**; the higher the number, the better. Minimum selectivity for an **audio** tuner or a receiver should be at least fifty decibels.

SEMICONDUCTOR.

A crystalline (often silicon) material whose electrical conductivity is somewhere between that of an insulator and a metal. Some semiconductor devices are transistors, diodes (including **light-emitting-diodes**), and solid-state lasers. **Integrated circuits** are semiconductor devices, also.

SEMICONDUCTOR MEMORY.

A **memory** which is made of a **semiconductor** material. Such memories usually are devised of "active" transistor circuits and often they must be **refresh**ed to retain their contents.

SENSITIVITY.

A specification applied to **tuners** and **receivers**. It indicates the smallest **signal** from the **antenna** that the device can convert into satisfactory sound or picture. Sensitivity, as a meaningful specification, should consist of two parts. The first is an expression of the electrical signal current in microvolts (i.e., millionths of a volt), frequently abbreviated μV; the second part, which indicates the limiting of the **noise** of the signal, is expressed in **decibel**s. The first number (that of the microvolts) should minimally be 2.5 or lower and the second figure (the decibels) should be 30 or higher. The Institute of High Fidelity (IHF)

SENSITIVITY (cont'd)
states that when sensitivity is given as an "IHF" specification for **audio** equipment, the microvolts figure will be given at a fixed noise limiting figure of 30 decibels. A specification reading simply "IHF sensitivity is 2.9 microvolts" is understood to mean that the missing portion is 30 decibels.

SENSITIVITY (PHOTOGRAPHY).
The degree to which an **emulsion** forms a **latent image** in response to specific light intensity levels and colors during **exposure**.

SEPARATION. *See* **channel separation**.

SERIAL ACCESS.
As opposed to **random access**, serial access requires the passing of all **data** or **information** on one side (or the other) of the specific data or information which is being sought. For instance, if the data or information is music stored on a **magnetic tape** and located in the center of that tape, one would have to pass (or search through) one or the other half of the tape to find what was wanted.

SERIAL TRANSMISSION.
The transmission of **data** or **information** in which the **character's bits** are transmitted one at a time, in sequence, over a single communication line. Contrasted with **parallel transmission** (occasionally called simultaneous transmission).

SERVO-CONTROL. *See* **servomechanism**.

SERVOMECHANISM.
A device, or group of devices, which uses **feedback** in order to determine such activities as the speed of a **servomotor** or the switching on of a heating system (the thermostat being a simple servomechanism). Servomechanisms are also called servo-controls.

SERVOMOTOR.
An electric motor whose speed is governed by a **servomechanism**. Good **tape recorders** and **turntables** are among those devices which employ servomotors.

SHARED FILE.
A **direct access storage** device which is used by two or more **computer** systems, and which may be used to link the systems logically so that **data** may be passed from one system to another.

SHEET FEED. *See* **document feed**.

SHELL. *See* **head shell**.

SHIBATA STYLUS.
A type of **stylus** for **phono cartridge**s especially designed (by a Japanese engineer named Shibata) for use with **CD-4 quadraphonic phonograph discs** designed for CD-4 type **quadraphonic** systems. The Shibata stylus is somewhat like a modified **elliptical stylus**; its advantage, however, consists of greater contact with the walls of the disc's grooves. The Shibata stylus tends to provide a better **frequency response** than the elliptical, particularly in the higher **frequency** part of the sound spectrum. The Shibata stylus, besides giving excellent quality with CD-4 discs, works exceedingly well with **stereophonic** ones.

SHIELDED CABLE. *See* **coaxial cable**.

146 / SHUTTER

SHUTTER.
   A mechanism that interrupts the light beam in a **film** camera or projector. In a camera, the closed shutter keeps the film from being exposed. In a projector it allows the film to progress from **frame** to frame while cutting off the light between frames. The most common camera shutters are the **between-the-lens shutter** and the **focal plane shutter**.

SHUTTER, BLADE. *See* **between-the-lens shutter**.

SHUTTER, FOCAL PLANE. *See* **focal plane shutter**.

SHUTTER SPEED.
   The speed with which a camera **shutter** opens and closes, the total time during which **exposure** of the **film** is allowed. Shutter speeds are usually stated in set, specific fractions of a second, e.g., 1/30, 1/60, 1/125, 1/250, 1/500, 1/1000. Few cameras offer shutter speeds faster than 1/1000 of a second. For shutter speeds slower than 1/125 of a second, it is advisable to use a **tripod**.

SIGNAL.
   The desirable **information** or intelligence conveyed in (or by) a communication system. The signal may take as its form a variety of energy types. For instance, the signal entering a radio is in the form of radio waves, but the same signal as it exits from the radio's **loudspeaker** is in the form of audible sound waves. The signal for the **video** portion of a television program is also in the form of radio waves, but it is ultimately converted into light waves, which emanate from the **cathode-ray tube**. **Noise** can be considered the opposite (or nemesis) of signal.

SIGNAL INPUT DEVICE (TELEVISION).
   Any device by means of which **images** and sounds are picked up and transmitted to the **videotape recorder** for recording, storage on **tape**, and reproduction. A **signal input** device may be a school district's television camera(s); other videotape recorders; or a broadcast station's **signal**s from its own cameras or television.

SIGNAL STRENGTH METER.
   A meter, frequently found on **tuner**s and **amplifier**s, which indicates the relative strength of the station to which the device is tuned. It is particularly helpful for tuning accurately to an **amplitude modulation** (AM) station. For **frequency modulation** (FM) stations, a **centering meter** is also of great use.

SIGNAL-TO-NOISE RATIO (S/N).
   A specification applied most often to **audio** equipment but occasionally also to **video hardware**. Usually expressed in **decibel**s, it is, simply, the ratio between the desired **signal** and undesirable **noise**. The higher the number, the better.
   In **tuner**s, **receiver**s, and **amplifier**s, a good signal-to-noise ratio would be sixty-five decibels, or higher. For **open-reel** type **tape recorder**s the figure should be minimally sixty decibels; for audio **cassette** machines, fifty-five decibels or better.

SILVER FILM.
   A **film** whose **emulsion** is composed of silver halide as the photosensitive material.

SILVER HALIDE FILM. *See* **silver film**.

SIMPLEX.
A **modem** design in which **data** are transmitted always in only one direction at a time. Contrasted with **duplex**.

SIMULATE.
The representation of a system or activity by some other means or system. For example, a **computer** may simulate manual bibliographic searching by means of **information retrieval**. Computers and other devices often simulate human activities or the activities of other devices, e.g., automobiles, aircraft, etc. Not to be confused with **emulate**.

SIMULCAST.
The simultaneous broadcasting of a television **program** (with its own **monaural** material) and the **audio** portion on a radio station (usually in **frequency modulation (FM) stereophonic** sound). The obvious purpose is to enhance the enjoyment of primarily musical programs like concerts and operas.

SIMULTANEOUS TRANSMISSION. *See* **parallel transmission**.

SINGLE-CONCEPT FILM. *See* **film**.

SINGLE LENS REFLEX CAMERA (SLR).
A camera in which **parallax** is eliminated, since the **viewing system** lens and the lens through which the **image** will be passed to the **film** are the same. SLR cameras are somewhat more expensive than those that use an **optical viewing** system.

SINK TERMINAL.
The receiving device in a **facsimile transmission** system.

16MM FILMS. *See* **film**.

SLAVES.
Magnetic tape recorders on which blank tape is run while the recorders are fed **signals** from a "master" tape on another recorder. This allows for the simultaneous **dubbing** of one tape onto a number of others.

SLIDE.
These come in a number of different sizes; however, the two types of slides that have been most widely used in schools are the 3¼x4-inch lantern slide and the 2x2-inch slide.

The lantern slide (one of the oldest devices used in visual education) is usually made from a 3¼x4-inch piece of thin glass. Its working surface (the side that has the **image** on it) may be photographic **emulsion**, ground glass, gelatin coat, or clear glass. In some instances special "slide crayons" can be used to draw an image or picture on the slide. Then the slide must be covered with a clear glass slide, and the edges bound with binding tape. In the past, **slide projector**s were mainly designed for 3¼x4-inch slides, but such projectors have been largely replaced by the 2x2-inch type, although some of these slide projectors can be adapted to utilize lantern slides.

The 2x2-inch slide (which is in predominant use now) consists of a piece of 35mm photographic **film** in a 2x2-inch cardboard, glass, or plastic mount. Color film is usually used for the 2x2-inch **transparency**, but it is possible to make black-and-white slides, if they

148 / SLIDE

SLIDE (cont'd)
are needed, for special purposes. Manufacturers can provide 2x2-inch slides in two sizes: half **frame** or full frame.

SLIDE EDITOR.
A backlit device which allows for the display and selection of **slide**s.

SLIDE FILM. *See* **color reversal film**.

SLIDE PROJECTOR.
Various types are available. Although they may differ in the sizes of the **slide**s they project, and in the way in which slides are transported and changed in the machine, they all operate pretty much on the same principle. Light from a **projection lamp**, with a reflector behind it, shines through the slide and the **image** is directed by the condenser lenses to the projection lens and then to the **screen**. The lens can be focused by turning the projection lens or by adjusting a focusing knob. In some machines focusing is achieved by pressing a remote control button and in others focusing is automatic.

Some slide projectors may have only a single slide carrier, in which one slide is placed and moved laterally to the **aperture**. The carrier can be permanently installed in the machine or detachable. In the latter case, the slide projector can sometimes be adapted to show **filmstrip**s. Multiple slide carrier machines can have rectangular slide trays, slide cubes, or rotary trays such as those designed for the Kodak Carousel Slide Projector.

Semi-automatic and automatic slide projectors all use trays or magazines for slide filing. The main difference between the semi-automatic and the automatic slide projector is the way in which the slides are changed. The former requires some action on the part of the operator, but the automatic type can be set to operate itself and change slides at a predetermined interval. An example of this type is the Carousel Automatic Slide Projector (Eastman Kodak Company).

SLOW-MOTION.
A **motion picture film** or **videotape** sequence that has been photographed (or recorded) at higher-than-normal speed; when the picture is projected (or played back on a **monitor**) at normal speed, the action on the **screen** appears much slower than the original action.

SMART TERMINAL.
A **terminal** which is not only an **input/output** device but which also contains its own **memory** (more than either an **intelligent terminal** or a **dumb terminal**) and is capable of maintaining a **program** in order to provide local processing.

SMOOTH SILVER SCREEN. *See* **screen**.

SNOW.
A speckly type of picture on a television set; it is almost always indicative of a weak **video signal**.

SOCKET. *See* **jack**.

SOFTWARE.
In terms of **instructional technology**, this refers to all those types of materials, such as pictures, photographs, **film**s, **filmstrip**s, **slide**s, **transparencies**, **program**s for teaching machines (for **programmed instruction**), **audio magnetic tape**, and **video tape** that are used with various audiovisual devices (**hardware**).

In **computer** parlance, software refers to programs.

SOLENOID.

An electromagnetic device which may, when coupled to a mechanical (or other type of) switch, activate, change the function of, or shut off a machine. Many **tape recorders** of better quality rely on solenoids rather than on simple **mechanical controls** only.

SOLID-STATE.

An electronic component that takes the place of other larger components like the vacuum tube. Solid state devices, e.g., transistors, diodes, etc., generate considerably less heat than their larger predecessors; but solid-state devices are more prone to failure caused by even relatively small amounts of heat. Good solid-state devices, when properly installed (i.e., installed in units that dissipate most of the heat to the air) last for a very long time and degenerate very slowly.

SOUND DRUM. *See* **motion picture projectors** and **optical sound track**.

SOUND HEAD. *See* **motion picture projectors** and **head** and **stripe**.

SOUND-ON-SOUND.

A method of **stereophonic** type **magnetic recording** in which the previously recorded material on one **track** is recorded onto another track while new material is simultaneously added to it. For instance, on track 1, there may be a previously recorded song played on a piano. Using the sound-on-sound method, that recording can be transferred to track 3 while there is added to it the playing of a guitarist, who records his part while listening to track 1 in order to synchronize his effort with that of the piano.

Sound-on-sound recording is sometimes called over-**dubbing**.

SOUND STRIPE. *See* **stripe**.

SOUND TRACK. *See* **optical sound track** and **stripe**.

SOUND-WITH-SOUND.

A method of **stereophonic** type **magnetic recording** which allows the user to listen to one **track** of a recording while recording himself on another track. It is frequently used in language laboratories; the instructor is heard on track 1, and the student can record his own pronunciation on track 3.

SOURCE MONITOR.

A way to **monitor** a **magnetic recording** device as it is recording. To use source monitoring, it is necessary to listen to the **preamplifier** or **recording amplifier** of the device—not to the actual **program** being recorded. Hearing the actual program being recorded, called **tape monitor**ing, is more useful, since it allows the listener to hear the finished product. Either type of monitoring, however, is better than no such capability, in that even source monitoring will provide some clue as to whether or not the device is recording properly.

SOURCE TERMINAL.

In a **facsimile transmission** system, the device which sends out the **facsimile**.

SPEAKER. *See* **loudspeaker**.

SPEECH SYNTHESIS.
A system used in some **computers** whereby human speech is **simulated**. Speech synthesis is most often encountered in systems which require statements (which may vary with time) to be made repetitively, as in flight departure times, banking transactions, etc.

SPHERICAL STYLUS. *See* **conical stylus**.

SPLICE, FILM.
The joining together of two pieces of **film**. In the case of a broken film, the two ends have to be cut off evenly or squarely: this is called a butt splice. For best results a suitably designed joiner or splicer should be used. With a cement splicer, the **emulsion** is removed from a narrow strip at the end of the film and an overlapped joint is made using a solvent cement. The joint is then held under pressure and heated. With a tape joiner, the squared ends of the two pieces of film are joined by transparent polyester tape. Never make a temporary splice with glue, gummed tape, etc., because film must be spliced only with special cement or special tape in a splicing machine.

A device specifically made for film splicing is often called a film editor.

SPLICE, TAPE. *See* **editing**.

SPLICE, VIDEOTAPE. *See* **editing**.

SPOOL. *See* **reel**.

SPOOLING.
A method in which **output** to devices slower than the **computer** itself is placed onto mass **storage** devices (e.g., **disk**s), which act as **buffer**s, and held in a **queue** to await final transmission.

SPROCKET.
A driving wheel with teeth around the outside rim. It is used to move **film** through various types of projectors by engaging with the film's perforations, which are called sprocket holes.

SQUAWKER. *See* **mid-range driver**.

STAND-ALONE.
Any piece of **hardware** which can operate with total independence of other devices.

STEP-AND-REPEAT CAMERA.
A camera used in creating **microform**s which can expose a series of individual **images** on a given section of **film** usually in columns and rows. Contrasted with the **planetary camera** and the **rotary camera**.

STEREO. *See* **stereophonic**.

STEREOPHONIC.
In **audio**, this describes a system (or a recording) that provides two **channel**s of sound separately and simultaneously. The two channels are usually designated left and right. Stereophonic recording of **phonograph disc**s began in 1958 in the United States, and it has become popular as a means of recreating, to a large extent, the **ambiance** of the environment in which a performance takes place.

STOP DOWN (PHOTOGRAPHY).

The reduction of the size of the **aperture** of a lens by one or more **lens stop**s.

STOP-MOTION.

The feature on some **motion picture projectors** or **videotape recorder**s, which allows the user to stop the advance of the **film** or tape through the device and to display a single **frame** of that tape or film. Some older projectors offered the stop motion feature but were improperly designed; if the film was stopped for more than a moment, the heat of the **projection lamp** burned the film. Newer projectors that offer stop-motion usually include a glass shield and a decrease in projection lamp brightness, which automatically protects the film (for a reasonable time) whenever the stop-motion function is engaged. Still, it is not wise to use stop-motion for longer than about a minute for film, or four minutes for tape (since the **helical head** in the videotape recorder will continue to rotate, and will cause wear to itself and the tape).

STORAGE.

A device which is capable of receiving and storing **digital data** or **information**. Frequently used synonymously with **memory**, storage is usually taken to differ in that storage may be remote from (and not online to) the **central processing unit (cpu)** while memory may be part of the cpu. Storage is also often used to mean a **medium** of mass storage like **magnetic tape**, **disk**s, etc., as opposed to memory **chip**s.

The term storage, when used by itself, might best be thought of as **auxiliary storage** as opposed to **main storage**.

STRENGTH METER. *See* **signal strength meter**.

STRIPE.

A magnetic **coating** applied to one or both margins of a **film** and used for recording sound. The sound can also be recorded optically on an **optical sound track**.

STROBE.

Properly termed a stroboscope, it is used to determine the accuracy of the speed of **turntable**s, **tape recorder**s, and **motion picture projectors**.

On turntables, a "strobe disc" (usually made of cardboard) can be placed on the **turntable platter** and a neon lamp held over the rotating disc. The dotted or lined pattern appropriate to the speed (most strobe discs are patterned for 33⅓, 45, and 78 rpm) at which the platter is rotating will appear to "stand still" if the speed is accurate.

Similarly, there are "strobe tapes" and "strobe films" for tape recorders and **film** projectors; these again, are used with a small neon light to determine the accuracy of the devices at various speeds.

STUDIO PRINT. *See* **matte**.

STYLUS.

Sometimes referred to as the "needle," this is a small, sharp point housed on a shaft (the cantilever) attached to a **phono cartridge**. The stylus retraces the undulations within a **phonograph disc** and imparts the mechanical vibrations to the phono cartridge, wherein they are converted into electrical impulses and fed to the **preamplifier, amplifier,** and thence to the **loudspeaker**s.

Most modern styluses are made either of sapphires or diamonds; the diamond stylus is the best in terms of length of use before wearing out.

STYLUS (cont'd)

There are now three types of styluses generally available: the **conical stylus, elliptical stylus,** and **Shibata stylus**. The first is useful for **monaural** long-playing discs, the second for **stereophonic** discs, and the third for stereophonic and **CD-4** discs.

A diamond stylus, if it is properly cared for (i.e., kept clean) and if its **tracking force** is appropriate, should have a useful life of from three hundred to eight hundred playing hours.

SUPER 8MM FILM. *See* **film**.

SUPERVISOR.

A **computer program** which is a part of, and which controls, a set of programs constituting a major computer application, or task. Such tasks might be word processing, **data base management**, etc. A supervisor is subordinate to an **executive program**.

SUPPLY REEL.

A spool in a **tape recorder** or **motion picture projector**. Tape or **film** is fed from the supply **reel**, through the device, and onto the **take-up reel**.

SWITCHED LINE.

A communication line between a **computer** and a remote **input/output** device whose connection is established by dialing (often referred to as **dial-up**). Contrasted with a **dedicated line**.

SYNCHRONIZATION.

Maintaining one operation or device in step with another. As it applies to television, see **sync-pulse**.

SYNCHRONOUS ORBITAL SATELLITE. *See* **communications satellite**.

SYNC-PULSE.

This term applies to the **synchronization** signals, or timing pulses, that lock the electron beam of the television picture **monitor** in step, both horizontally and vertically, with the electron beam of the **pickup tube** of the television camera. Synchronization is also an important factor in **filmstrip**-record, **cassette**-filmstrip or **slide**, and slide-tape systems in which the pulse or signal has to project the proper **frame** or picture upon the **screen** at the right time.

SYNTHESIZED FOUR CHANNEL. *See* **derived four channel**.

SYSTEMS ANALYSIS.

The examination of a business, operation, function, etc., in order to determine both what objectives are to be accomplished and how best to accomplish those objectives. Not to be confused with **operations research**. Systems analysis often begins with a **flowchart** in order to conceptualize better the operation, function, etc.

SYSTEMS APPROACH.

A process used to examine all the **input**s of an instructional **program** (such as objectives, available resources, and personnel) in an attempt to evaluate its effectiveness and to resolve problems.

THD. *See* **harmonic distortion**.

TTY. *See* **teletypewriter**.

TVR. *See* **kinescope recording**.

TWX.
   A **teletypewriter** communication system used in **real time**. In other words, a system of communications through a specific **network** using **keyboard** style **terminal**s (and, occasionally, **paper tape** or other **input** means); the **output** is a **printout** at a remote station. TWX, in so far as the user is concerned, is much like a typing telephone system.

TACHISTOSCOPE.
   A variation of the **slide projector**, produced by the addition of a special type of **shutter**. The shutter attachment allows the operator to flash numerals, words, sentences, and paragraphs (or pictures) on a **screen** for times varying from 1 second to 1/100 second. The tachistoscope has had limited use in the teaching of spelling, number combinations, and certain phases of reading.

TACKING IRON.
   A small, electrically heated device (much like a tiny iron used for pressing clothes, except that it has a long handle). It is used to heat and melt (in the **dry mounting of pictures**, maps, etc.) the tissue placed between a picture and a cardboard (or cloth) background.

TAG. *See* **flag**.

TAKE-UP REEL.
   A spool in a **tape recorder** or **motion picture projector** upon which tape or **film** is wound after having been fed from the **supply reel** and through the device.

TAPE. *See* **magnetic tape**.

TAPE CARTRIDGE. *See* **cartridge, audio**.

TAPE CASSETTE. *See* **cassette, audio**.

TAPE DECK. *See* **deck**.

TAPE HEAD. *See* **head**.

TAPE HISS.
   The **noise** that is peculiar to **audio** type **magnetic tape** recording in all its forms, e.g., **open-reel**, **cassette**, etc. Technically described as random, high **frequency** noise, it takes the form of a kind of sibilance that is most often discernible during quiet recorded passages. The slower the **tape speed**, the greater the tape hiss, particularly in formats that have narrow **track**s and/or slow tape speeds like cassettes and **cartridge**s. In order to reduce the amount of tape hiss, various noise reduction systems have been introduced in tape recorders, the most popular being the **Dolby noise reduction** system.

TAPE INPUT.

This is the **jack** on an **amplifier, preamplifier,** or **receiver** that accepts the **output signal** produced by a **magnetic tape** playback or recording device, in order to feed that signal to the rest of the **audio** system. The output jack of the magnetic tape device is connected to the tape **input** jack by means of a **patch cord**.

TAPE LEADER. *See* **leader**.

TAPE LIFTERS.

Mechanical devices in better **open-reel** type **tape recorder**s that lift the **magnetic tape** away from the **head**s when the tape recorder is in the **fast forward** or **rewind mode**.

Tape lifters prevent unnecessary contact of heads and tape in these modes; thus, the friction created when the tape is running at these fairly high speeds does not wear the heads. Further, tape lifters lessen the amount of "Donald Duck" **noise** one hears as the tape is being run very quickly.

TAPE MONITOR.

A way to **monitor** a **magnetic recording** device as it is recording. As opposed to **source monitor**, the tape monitor function allows the user to monitor the actual finished recorded tape a fraction of a second after the recording has been made. Tape monitoring is the most useful method of monitoring, since the recordist can adjust the recording process most accurately.

TAPE OUT.

The **jack** on an **amplifier, preamplifier,** or **receiver** which feeds a **signal** from one portion of the **audio** system to the **input** jack of a **magnetic recording** device in order to record that signal. For instance, a program derived from a **disc** played on a **phonograph** (which is hooked up to an amplifier) may be recorded on an audio **cassette** recorder (which is also connected to the amplifier with **patch cord**s).

TAPE RECORDER.

A device that can record either **audio** or **video** material, or both, on **magnetic tape**.

An audio tape recorder must contain an **erase head**, a **record head**, usually a **playback head**, a **tape transport**, and a **recording amplifier**. There should also be a **digital counter** and some metering system that indicates recording (and possibly playback) **amplitude**.

A **videotape recorder** must include all the items mentioned above, plus a **video head**.

Portable tape recorders, whether audio or videotape, include **loudspeaker**s and usually (for the videotape device) **monitor**s.

Audio tape recorders come, especially in three forms: **open-reel, cartridge,** and **cassette**.

Videotape recorders are ordinarily available as **open-reel** or **video cassette**.

Specifications applicable to tape recorders are **channel separation, crosstalk, frequency response, harmonic distortion, intermodulation distortion, signal-to-noise ratio, wow,** and **flutter**.

TAPE SPEED.

This is the speed at which **magnetic tape** in a **tape recorder** moves past a fixed point. Tape speed is usually specified in **inches-per-second** (ips).

In most situations, the faster the tape speed, the better the quality of the recording.

Audio tape recorders are usually provided with any (or a combination) of four possible specific speeds: 15 inches-per-second, 7½ inches-per-second, 3¾ inches-per-

TAPE SPEED (cont'd)

second, and 1⅞ inches-per-second. All these speeds are to be found in various combinations in **open-reel** devices; the second two are the most common, and the last three are not unusual. **Cartridge** machines operate at 3¾ inches-per-second only, and **cassette** at only 1⅞ inches-per-second. **Videotape recorders** vary widely in their tape speeds.

TAPE THICKNESS. *See* **base** and **mil**.

TAPE TRANSPORT.

The mechanical system of a **tape recorder** which moves the tape from the **supply reel** to the **take-up reel**, and vice versa. The transport consists of the motive force (usually a motor which transmits rotation by means of belts; or three motors, one for the record and playback **modes**, one for **fast-forward**, and one for **rewind**), the **capstan**, **head**s, tape guides, controls, and spindles onto which the **reel**s are placed. The tape transport does not include the electronic components such as the recording **amplifier**.

The three-motor transports are better, in that **tape speed** tends to be more uniform and **wow** and **flutter** are minimized.

TAPE TYPE.

There are now four standard **magnetic tape** types which are available for use primarily in **audio**-style **tape recorder**s:

Type I is **ferric oxide tape**.
Type II is **chromium dioxide tape**.
Type III is **ferrichrome tape**.
Type IV is **metal tape**.

The different types of tape require some difference in **bias**; generally, however, in playback, those devices which can play type II can also play type IV, since both require a 70 microsecond **equalization**, while type I requires a 120 microsecond equalization.

Type III, as discussed under the entry for ferrichrome tape, has fallen into disuse and very few machines are now available which can properly utilize it.

TARGET.

Properly a **resolution** target, a target is a matrix of lines (usually at right angles to each other) which are of diminishing sizes. The target is usually found at the beginning of a **microform** and acts as an indicator of the resolution quality of that microform or copies made from it.

TEACHING FILM. *See* **film**.

TEACHING MACHINES. *See* **programmed instruction**.

TELE-FAX. *See* **facsimile transmission**.

TELENET.

The common name for the Telenet Communications Corporation, an American company which provides less expensive long distance rates to remote **computer**s than does the American Telephone and Telegraph Company.

TELEPHONE ANSWERING DEVICE.

A piece of **hardware** (consisting of at least one **tape recorder**) which can be connected to a telephone line. The tape recorder plays a prerecorded message to callers when the

156 / TELEPHONE ANSWERING DEVICE

TELEPHONE ANSWERING DEVICE (cont'd)
phone is unattended. More expensive models incorporate a second tape recorder which records the callers' messages. Some of these devices can be activated remotely, that is the owner may have the device play back the callers' messages when the owner dials into the system and activates the machine with a special device.

TELEPHOTO LENS. *See* **focal length**.

TELEPRINTER. *See* **teletypewriter**.

TELEPROCESSING.
Any **data processing** system which can be used via some communication system so that the user may be quite remote from the **computer**.

TELETEXT.
Teletext (and viewdata) are generic terms used to describe the broadcasting of additional **data** and **information** (both in textual and **graphic** form) as part of a regular broadcast television **signal**. The television **receiver** must be equipped with a special decoding device in order to make the additional materials viewable on the regular television **raster**.

TELETYPEWRITER.
A **terminal** which includes a **keyboard** and a **printer** and which is used to communicate with a **computer** or to transmit **data** in a communications system.

TELEVISION BROADCAST.
A transmission, live or recorded, of pictures and sounds in electronic form, made by a station licensed by the **Federal Communications Commission** (FCC) for such a purpose. It is sent through space from the point of origin to the television **receiver** without any physical connection in between.

TELEVISION CAMERA. *See* **pickup tube**.

TELEX.
An international **computer** communications system which, in the United States, is provided by Western Union.

TENSILIZED TAPE. *See* **pretensilized tape**.

TENSION CONTROL.
A system found in most better **tape recorder**s. It maintains equalized tension of the tape in the record and playback **mode**s.

TERMINAL.
An **input/output** device which may consist of a **keyboard, printer, cathode-ray tube, modem** or any combination thereof.

TEST PRINT. *See* **contact print**.

THERMAL PRINTER.

A **printer** which, unlike an **impact printer,** uses a specially coated heat sensitive paper on which the **characters** are formed by the action of the printing mechanism (which is somewhat warm). The characters formed by a thermal printer are of the **dot matrix printer** type. Thermal printers, although generally slower than impact printers, tend to operate rather more quietly.

THERMOGRAPHIC COPYING.

A **copying machine** technique which is based on the principle that dark colors absorb more heat than light ones. Thermographic copiers, therefore, use a specially coated (heat sensitive), translucent paper. Within the device an **exposure** of the copy paper is made while in contact with the document to be copied. The exposure is done under infrared light. The thermographic copy, due to the nature of the paper used and the process itself, is a flimsy one which is unstable and tends to fade. The real advantage to thermal copying over **electrostatic copying** is the relatively low cost of the equipment. It is also, however, slow and not intended for high volume copying operations.

THIN FILM MEMORY.

A **medium** of **storage** composed of an extremely thin (i.e., close to a single molecule in thickness) magnetic material which has been deposited on an insulating material while in a partial vacuum.

THREADING, FILM. *See* **motion picture projectors.**

THROUGHPUT.

The productivity, as against time, of a system, from start to finish. For instance, an assembly line which begins by mixing the ingredients for bread, and continues through the baking, slicing, and wrapping stages may have a throughput of five hundred loaves an hour. A similar, but either slower or less productive system, would be said to have low (or lower) throughput.

TILT-PAN HEAD. *See* **tripod.**

TIME-LAPSE PHOTOGRAPHY.

A photographic system (usually employing motion picture equipment) which makes an **exposure** of one **frame** at a time at regular, preset intervals. When either the still or motion pictures are viewed, the process or changes in the subject appear to have occurred relatively quickly or in a short time. Time lapse photography is often used to depict such slow processes as the opening of a flower, etc.

TIME SHARE.

The means by which two or more uses may be made of the same system, **computer**, etc., concurrently. This is accomplished by having the system work on each use for a very short period of time and then go on to the next. If the system is sufficiently rapid, it would appear that only one use (or user) is being accommodated rather than several.

TONE ARM.

The part of a **phonograph** into which the **phono cartridge** is fitted; it is usually mounted on the **turntable.** The tone arm is fixed at one end to a pivot, which allows it to swing in an approximation of a radius across the surface of the **phonograph disc.** In order not to be too heavy, the **tracking force** (or weight) of the tone arm should be adjustable so

TONE ARM (cont'd)

that it can accommodate phono cartridges of different weights. The tracking force adjustment is made by means of a counterweight in good tone arms or by means of spring loading in cheaper ones. In addition, good tone arms usually have some kind of **antiskating** arrangement. Another type of tone arm is the **radial arm**.

One important specification of a tone arm is the measure of its **tracking error**; another is its **resonance frequency**.

TONE CONTROLS.

These are controls found on **amplifiers**, **preamplifiers**, or **receivers**; usually they adjust the **bass** or **treble output** of the device. In **stereophonic** units, it is helpful to have the left and right **channel**s controlled separately, instead of having one knob that controls the treble (or bass) for both channels.

The tone controls in most devices are continuous and nondiscriminatory, that is, nonselective. To reduce **tape hiss**, which may occur at twelve thousand **hertz**, all frequencies above twelve thousand hertz must also be reduced, thereby causing a lessening of brightness of the **program**. A solution to this problem is the use of a **frequency balance control**, which is simply a very selective and sophisticated tone control.

TONER.

The **medium** in an **electrostatic copying** machine or **electrofax copying** machine which ultimately becomes the **image** on the copy. There are both dry toners and wet toners. The former are most peculiar to electrostatic copying while the latter are almost always used in electrofax copying.

TONE GENERATED DIAL.

Tone generated dial (frequently referred to as touch tone dial) is the method of sending a **signal** in a telephone system in order to place a call or "talk" to a **computer** or other **data processing** equipment. A tone generated dial telephone has buttons which, when depressed, create specific audible tones, each of which represents one **character**. Contrasted with **rotary dial**. Touch-Tone is a registered trademark of the American Telephone and Telegraph Company.

TOOTHED BELT.

In a **belt drive** system, a toothed belt is one which has gear-like teeth on its driving surface, this as opposed to the more typically found smooth surface belt. The teeth mesh with similarly sized indentations on the surface of the pullies on the motor and the driven device. Toothed belt systems are most often found in machines like **motion picture projectors**. Proponents of the toothed belt claim that it prevents the potential slippage of a smooth belt.

TOTAL HARMONIC DISTORTION. *See* **harmonic distortion**.

TOUCH-TONE DIAL. *See* **tone generated dial**.

TRACK.

The path of reproducible **information** on **magnetic tape**. For instance, in a **stereophonic** magnetic tape recording, the left **channel** actually consists of a physically magnetized path that runs the length of the tape. In a stereophonic audio **cassette**, this path would be designated (on side one of the cassette) as track 1; on side two, it would be designated as track 3; etc.

TRACKABILITY.

The ability of a **stylus** in a **phono cartridge** to retrace **phonograph disc** grooves that are of both high **frequency** and high **amplitude**. Trackability, as a specification, is usually cited as **compliance**.

TRACKING ERROR.

A specification that indicates the amount by which a **phono cartridge** deviates from perfect tangency with the **phonograph disc** at a given point—that is, the amount that it deviates from tracing a perfect radius across the disc's surface. Tracking error is given in degrees-per-inch and should be less than two or three degrees.

TRACKING FORCE.

The amount of downward thrust (or weight) a **phono cartridge** requires in order for its **stylus** to retrace properly the grooves in a **phonograph disc**. Tracking force is always expressed in grams. In order to minimize both disc and stylus wear, a lighter tracking force is desirable. The optimal range is from approximately .5 grams to 1.5 grams.

TRANSCEIVER.

A **facsimile transmission** device which can act as either a transmitter (or sending **terminal**) or a **sink terminal**.

TRANSDUCER.

A device capable of converting one form of energy to a different form. For instance, a **loudspeaker** converts electrical energy into acoustical energy, a **microphone** converts acoustic energy into electrical energy, and a **phono cartridge** converts mechanical energy into electrical energy.

TRANSIENT.

A sudden, brief (and usually high amplitude) **signal**. In **audio**, transients take such forms as a cymbal clash, a tympani stroke, or a pizzicato on the violin.

TRANSIENT NOISE REDUCTION.

The reduction of **noise** of a **transient** type in a sound system. Such noise is usually caused by static electricity, dirt, or scratches on a **phonograph disc** surface and may sound like ticks, clicks, or pops. Transient noise can also occur on **magnetic tape** and is the result of **dubbing** from a scratched or dirty phonograph disc, or may be a function of tape **dropout**.

In any event, there are special devices which will reduce or eliminate transient noises by any of a variety of means.

TRANSIENT RESPONSE.

The ability of a device to reproduce a **transient** accurately. Transients cause an abrupt **signal** pulsation and what should be an abrupt decay of that signal; thus, there should be little or no ringing or after image produced. Cymbals should clash and not sound sandpapery, pizzicati should sound almost palpable, etc.

TRANSLATOR.

An electronic device capable of receiving a television **signal** from one source of transmission or **channel** and retransmitting it on another channel. It does this by means of direct **frequency** conversion and amplification of the incoming signals without significantly altering any characteristic of the incoming signal, except its frequency and **amplitude**.

TRANSPARENCY.

A transparent sheet of plastic (acetate) with printed or diagrammatic material on it, which is displayed on a **screen** by means of an **overhead projector**. It may or may not be mounted in a cardboard surround. Commercially prepared transparencies (in black-and-white or color) cover almost every subject in the curriculum and all grade levels. (The **National Information Center for Educational Media** [NICEM] provides an index of educational overhead transparencies). Some transparencies use **polarized projection** to create the illusion of motion. Other types use an overlay technique, in which the first sheet placed on the overhead projector displays basic **information**, and subsequent sheets, containing more information, are flipped over in sequence to build upon the base.

The teacher can write directly on blank transparency sheets or rolls with a felt pen or grease pencil. Also, transparency information can be typed on a special clear acetate called Type-On, which can be placed directly in the typewriter.

Several techniques can be used to produce transparencies, but they require intermediate materials and different types of machines. Heat-process transparencies can be made from original materials by using a Thermo-Fax copy machine or a Master Fax unit. Spirit-duplicator masters can be used to make transparencies by running a sheet of finely etched acetate through the duplicating machine. Transparencies can also be made by the photocopy process and the electrostatic process (Xerox and A. B. Dick), both of which involve creating paper duplicates and then an acetate transparency for projection.

Diazo film transparencies (ammonia process) are made by first preparing the master copy on high quality tracing paper or translucent acetate using India ink. Next, a sheet of diazo film is placed against the drawing on the master copy and the two are laid in an **ultraviolet**-light exposing unit with the tracing paper or acetate first in position to the light and the film next. **Exposure** time is critical, and the directions that accompany the film or diazo machine should be followed exactly. Finally, the exposed sheet of diazo film is developed in a large jar of ammonia vapor or in the diazo machine.

The picture transfer, or lift process, can also be used to make transparencies from pictures printed on clay-coated paper. To test whether the paper is clay-coated, rub a dampened finger on the page margin. If a white, chalky substance comes away on the finger, the picture can be lifted (but the page will be destroyed in the process). Cover the picture with a clear, adhesive-backed shelf paper such as Con-Tac. It should be firmly pressed onto the picture. The picture and the shelf paper should then be soaked in cool water for a few minutes. The shelf paper with the picture on it can then be peeled away from the paper it was originally on. Any clay residue should be washed away, and then the transparency can be dried and sprayed with a clear plastic to protect it. It can be mounted in a cardboard surround like any other transparency.

TRANSPONDER.

In a communications system, a transponder is a special device which can accept certain types of inquiries and automatically provide the correct responses.

TRANSPORT. *See* **tape transport**.

TREBLE.

The sounds in the higher **audible frequency range**, usually from about 3,000 **hertz** to about 18,000 or 20,000 hertz. For the sake of comparison, the note C, three octaves above middle C (256 hertz), has a frequency of 2,048 hertz, while the C an octave above that (which is the highest note on many pianos) is 4,096 hertz.

TREBLE (cont'd)

Treble sounds tend to be very directional—that is, their source of emanation is considerably more obvious than is the source of **bass** sounds.

The **loudspeaker** component usually associated with the reproduction of treble sounds is the **tweeter**.

TRIPOD (PHOTOGRAPHY).

A three-legged stand on top of which is a camera mount. The tripod ensures greater steadiness to the camera when making **exposure**s of relatively slow **shutter speed**s. Better tripods are equipped with tilt-pan heads which allow the camera mount to be moved through a variety of planes. A pan head is a camera mount which moves 360 degrees, but only parallel to the ground (for taking panoramic photographs or motion pictures).

TRUNCATE.

In **data bank** material **retrieval**, the shortening of a search term to its most abbreviated, yet specifically identifiable, form in order to retrieve all possible forms of that term based upon the same root form. For example, the truncated form "manag-" should retrieve all material indexed under the terms "manageable," "management," "manager," "managerial," "managing," etc. Obviously great care should be exercised in using the concept of truncation.

TUNER.

A device that is capable of receiving and selecting either radio or **television broadcast**s. A digital tuner or receiver has a **digital display**.

Audio tuners are usually designed to receive and select **amplitude modulation (AM)**, and/or **frequency modulation (FM)**, or just FM or FM **stereophonic** broadcasts. Such audio tuners must be hooked up to an amplification system and **loudspeaker**s, since the tuner itself contains none of these.

Specifications appropriate to audio tuners are **capture ratio, channel separation, frequency response, harmonic distortion, image rejection, selectivity, sensitivity,** and **signal-to-noise ratio.**

A tuner designed to receive television broadcasts is usually an integral part of a television set or **monitor**; its specifications are usually given as functions of the entire device.

TUNER INPUT.

The **jack** on an **amplifier, preamplifier,** or **receiver** which is intended to accept the **output signal** of a **tuner**. Usually this output signal has a voltage of about .5 volts, and the design of the device into which this signal is fed should accommodate that figure.

The tuner's output jack is usually connected to the amplifier (or whatever device is used) by means of a **patch cord**.

TURNAROUND TIME.

The time it takes from the submission of a job to be done until its completion.

TURNKEY.

A device or system whose supplier assumes all responsibility for construction, installation, and **maintenance**. The owner need only "turn the key" to set the device or system into operation. Of course, the turnkey system, much like the automobile, requires that the owner know how to manipulate the system once it is operating.

TURNTABLE.

The part of a **phonograph** that contains the **tone arm** (and **phono cartridge**, in the tone arm), **turntable platter**, and the motor to drive the turntable platter.

The better turntables use either a **direct drive** or **belt drive** system, while cheaper ones use the rim **drive** type. Better turntables are designed to minimize **acoustic feedback**, usually by some sort of shock-absorbing system.

The turntable should rotate the **phonograph disc** at an accurate (within .2 percent) speed (usually 33⅓ or 45 rpm) and provide a good **tone arm** or a place where a tone arm can be mounted.

Better turntables use **hysteresis motors**; less expensive ones use induction motors. The speed accuracy of a turntable can be determined by means of a **strobe** disc.

Because of various complications, the automatic record changer (or automatic turntable) — which automatically places a disc (from a number of discs loaded on its spindle) onto the turntable platter and automatically positions the tone arm on the disc, etc. — is definitely not recommended for institutional use (or for any other use, for that matter). Such a device does not handle discs properly, and it does not withstand constant use.

Important specifications for turntables that are not equipped with a tone arm are **flutter, rumble,** and **wow**.

For turntables equipped with tone arms and phono cartridges, the important specifications, in addition to the above, are **channel separation, frequency response, resonance frequency, tracking error,** and **tracking force**.

TURNTABLE PLATTER.

That part of a **turntable** on which the **phonograph disc** rests and is rotated. In general, better turntables use heavy platters (over two pounds), and some have **strobe** discs incorporated into them.

TURRET, LENS.

A rotating plate that is mounted on the front of a camera and that holds two or more lenses. By rotating the turret, the operator can rapidly bring any one of the lenses into position for use.

TWEETER.

A component of many **loudspeaker** systems. It is in itself a loudspeaker, but it is designed to reproduce the upper portion **(treble)** of the **audible frequency range** from about three thousand **hertz** to about eighteen thousand hertz. Frequencies below this range are ordinarily reproduced by loudspeakers termed **mid-range driver** and **woofer**.

TWIN LENS REFLEX CAMERA. *See* **reflex camera.**

TWO TRACK. *See* **half track.**

UHF. *See* **ultra high frequency.**

UL. *See* **Underwriters Laboratories.**

UPC. *See* **bar code.**

USASCII. *See* **ASCII.**

USASI. *See* **American National Standards Institute.**

UV. *See* **ultraviolet**.

ULTRA HIGH FREQUENCY (UHF).
For television, the UHF **band** extends from 470,000,000 **hertz** to 890,000,000 hertz and covers **channels** 14-83. The UHF band is not as powerful as the **very high frequency** (VHF) band. This means that the UHF band does not cover as large a broadcast reception area as the VHF band.

ULTRAFICHE. *See* **photo-chromic-micro-image**.

ULTRAVIOLET.
That light which begins at the short wavelength end of the visible spectrum and continues into the long x-ray wavelengths.

ULTRAVIOLET FILTER (UV FILTER).
A lens **filter** (whose **filter factor** is zero) which keeps ultraviolet light from being part of the light spectrum which is available for **exposure** of the **film**. The UV filter makes skies darker (or bluer) and clouds whiter and more distinct.

UNDEREXPOSE (PHOTOGRAPHY).
To prevent sufficient **exposure** of photosensitive material (e.g., **film, print** paper, etc.) because of insufficient light, too brief an exposure time, or an **aperture** setting which is too small. Underexposed pictures tend to look dark and indistinct and without much (or, contrariwise, often too much) contrast in tonal range.

UNDERWRITERS LABORATORIES.
A testing laboratory established in 1893 to evaluate various types of equipment and their uses with respect to hazards. Laboratory divisions are Burglary Protection; Casualty and Chemical Hazard; Electrical; Fire Protection; Hazardous Location; Heating, Air-Conditioning and Refrigeration; and Marine. Annual Lists of tested devices, materials, and methods are published by UL, headquarters of which is located at 333 Pfingsten Rd., Northbrook, IL 60062.

UNIDIRECTIONAL MICROPHONE. *See* **cardioid pattern**.

UNITED STATES OF AMERICA STANDARDS INSTITUTE. *See* **American National Standards Institute**.

UPDATE.
The addition to a **data bank, memory,** system, etc., of new material (including additions, deletions, changes, etc.) so that currency is reflected.

USER FRIENDLY.
A **computer** system, usually employing the techniques of an **interactive system**, which leads the user to desired **data** or **information** by providing instructions, help, options, **prompt**s, etc., in order to obviate the user's need to use a manual or assistance from some **interface** (usually human).

USER IDENTIFICATION.
A set of **characters** which uniquely identifies a specific user to a **data processing** system. User identifications, like **password**s are usually considered confidential. The user

USER IDENTIFICATION (cont'd)

identification, however, is essentially used to determine for purposes of accounting how much time (and, often, what portions of the system) a user has used.

VCR. *See* **video cassette.**

VDT. *See* **video display terminal.**

VHF. *See* **very high frequency.**

VHS.

The abbreviation VHS stands for Video Home System and is a **video tape** style **tape recorder** format similar to, but not compatible with the Sony **Betamax** system. VHS was developed by the Victor Company of Japan (JVC) to compete with Betamax, and, at the time of this writing, has done so with remarkable success. In the United States, VHS systems are licensed to, and sold by, such companies as Magnavox, RCA, General Electric, and Panasonic. The cassettes use a tape which is half an inch wide and has a playing time which varies from one-half hour to eight hours. VHS cassettes are designated by their playing time (in minutes) at the fastest playback speed. For example a T30 cassette will play thirty minutes at the fastest speed, or ninety minutes at the slowest; the speed capabilities being a function of the specific VHS model. There is essentially no quality difference between VHS and Betamax formats.

VTR. *See* **videotape recorder.**

VU GRAPH. *See* **view graph.**

VU METER.

Properly called the volume unit meter, this is standard equipment on many **tape recorders**. It is used to indicate the **amplitude** (volume) of the **signal** being recorded (or played back).

This type of meter is designed to indicate amplitude in **decibel** units; "0 VU" is a reference power of one milliwatt in a circuit of six hundred ohms characteristic **impedance**. Numbers above or below "0 VU" indicate the decibel count above or below that reference.

In many tape recorders, the recording **level** control should be set so that the loudest portions of the program are slightly above "0 VU"—say, to plus one or two decibels. In devices that have built-in **Dolby noise reduction** system circuits, the setting can be higher—say, to plus two or three decibels VU. Now, **light-emitting-diode** or **fluorescent light** displays have effectively replaced the VU meter.

VARIABLE LENGTH.

As opposed to **fixed length**, variable length describes a group of **characters** constituting a **data** record or **field** which has essentially no limit on the number of those characters.

VELCROBOARD. *See* **clothboard.**

VENN DIAGRAM.

A logical diagram depicting mathematical, or logical, sets as circles which have varying relationships depending on their logical or mathematical connections—e.g., **Boolean operators** are easily represented by Venn diagrams.

VENTED BAFFLE. *See* **baffle.**

VERY HIGH FREQUENCY (VHF).
For television, the VHF **band** extends from 44,000,000 **hertz** to 88,000,000 hertz for **channel**s 2-6 and from 174,000,000 hertz to 216,000,000 hertz for channels 7-13. The VHF band is stronger than the **ultra high frequency** band; therefore, it covers a larger broadcast reception area.

VESICULAR FILM.
A **film** which does not have an **emulsion,** in the usual sense (as, for example, **silver film**), but, rather, consists of a light sensitive material (usually in a gaseous form) which is trapped on the surface of the film in microscopic bubbles (vesicles). The film, on **exposure,** does not require further effort to adduce the **latent image** because the act of exposing it (to heat and light, usually **ultraviolet**) also **develop**s it. The most commonly known vesicular film is Kalvar.

VIDEO.
This pertains to television or **film**; it designates the picture portion of the **program** as distinguished from the **audio** (or sound) portion.

VIDEO AMPLIFIER. *See* **amplifier.**

VIDEO CASSETTE.
A plastic container that houses **videotape** (as opposed to **open-reel**) and that can be inserted as a unit into the video cassette machine.

VIDEO DISK. *See* **optical disk.**

VIDEO DISPLAY TERMINAL (VDT).
A **terminal** which incorporates a **cathode-ray tube** display to depict **input** and/or **output**. Most commonly, VDT units also have a **keyboard** and may use a **printer.**

VIDEO-GRADE CHANNEL.
A communications **channel** of a sufficient **frequency range** to allow the transmission of **video image**s. A video-grade channel must be of considerably greater frequency range than a **voice-grade channel.**

VIDEO HEAD.
The **head** in a **videotape** or **video cassette** device which can record and/or play back the visual **information** on **magnetic tape**. Most video heads are of the **helical head** type.

VIDEO HOME SYSTEM. *See* **VHS.**

VIDEO MONITOR. *See* **monitor.**

VIDEOTAPE.
A **magnetic tape** on which **image**s (**video**) and sound (**audio**) are recorded and reproduced by the **videotape recorder**. It is very much like audio tape. Videotape varies in width from one-quarter-inch to two-inches and is usually at least one **mil** in thickness. Videotape can be recorded either in black-and-white or color.

166 / VIDEOTAPE RECORDER

VIDEOTAPE RECORDER.
An electromechanical device that makes possible the electronic recording and immediate playback of television **images (video)** and sound **(audio)** on **magnetic tape**. Most **videotape** devices can both record and play back, but some are designed to play back only. Videotape machines can either be **open-reel** devices or can accept **video cassette**s.

The videotape recorder accepts the **signal**s from the **output** of either a camera and **microphone** and/or a television set, or another videotape recorder.

Important specifications for videotape devices are **frequency response, resolving power, signal-to-noise ratio, wow,** and **flutter**.

VIDEO TUBE. See **cathode-ray tube**.

VIDICON CAMERA TUBE.
The vidicon camera is a television camera which is sometimes known as an industrial camera. The vidicon camera, with simpler external circuitry than the **orthicon image camera tube**, has made possible smaller and less expensive cameras. It is advantageous for educational and industrial applications because of its lower operating costs. The selection of a **view finder camera** as opposed to an industrial camera should be based on use requirements. In those situations where on-site operator control is required, an industrial model is almost totally unacceptable. However, the use of a view finder camera where an industrial model would suffice holds considerable implications in terms of cost.

VIEWEDATA. See **Teletext**.

VIEWER. See **hand viewer**.

VIEW FINDER. See **viewing system**.

VIEW FINDER CAMERA. See **range finder camera**.

VIEW FINDER CAMERA (TELEVISION).
A television camera with a small viewing **monitor** mounted on it, or built into it. It is usually found with **orthicon image camera tube**s, but may also be used with the **vidicon camera tube**.

VIEW GRAPH.
A **transparency** containing graphs, charts, etc., intended for use with an **overhead projector**.

VIEWING MONITOR. See **monitor**.

VIEWING SYSTEM.
A generic term indicating the portion of a camera through which the user views the subject to be photographed. There are two basic types of viewing systems: **optical viewing system** (also called auxiliary or external) and the **reflex viewing system**. Of the two, the reflex viewing system is the better; the **reflex camera** is somewhat easier to use, and the **single lens reflex camera** eliminates **parallax**.

VIRTUAL IMAGE.
An apparent optical **image** which seems to emanate diverging light rays, but without actually being focused at the apparent site. An example is the image that appears to be behind an ordinary mirror.

VIRTUAL MEMORY.
A **computer** system which employs **auxiliary storage** (usually **disk**) in such a way that the user is allowed to treat that storage as if it were **main storage**. The auxiliary storage is, in effect, "transparent" to the user.

VOICE COIL. *See* **piston speaker**.

VOICE-GRADE CHANNEL.
A communications **channel** capable of transmitting the **frequency range** required to encompass the sound of human speech (from about three hundred **hertz** to thirty-four hundred hertz). A regular telephone line is one such channel. Contrasted with **video-grade channel**.

VOLATILE MEMORY.
A type of **storage** which, when its energy source is denied to it, loses all stored **data** or **information**.

VOLUME. *See* **amplitude**.

VOLUME CONTROL.
Sometimes called the **gain** control or **level** control, this is the control on an **amplifier** that can increase or decrease the **amplitude** (or volume) of the **output**—that is, in **audio**, make the sound louder or softer. In many devices, the volume control incorporates the power off/on switch.

VOLUME INDICATOR. *See* **VU meter**.

WAND. *See* **light pen**.

WATS. *See* **Wide Area Telephone Service**.

WAVELENGTH.
The measurement, usually in meters, of the wave of a given **frequency**, as measured from peak to peak or trough to trough.

WEB FEED.
Copying machine paper stored on a large roll (as opposed to a stack of single sheets) and cut to size as it is used.

WET CARREL. *See* **carrel**.

WET TONER. *See* **toner**.

WHITE NOISE.
A random **noise** comprised not of one **frequency**, but of a multiplicity of frequencies. In **audio**, white noise sounds much like static.

WHOLE TRACK. *See* **full track**.

WIDE ANGLE LENS. *See* **focal length**.

WIDE AREA TELEPHONE SERVICE (WATS).

A telephone service of the American Telephone and Telegraph Company in which the subscriber leases a line, for direct-dialing purposes, on a monthly, flat fee basis. The United States is divided into six regions in order to determine rental rates. WATS may be rented to allow incoming calls only or both incoming and outgoing calls. The latter, of course, carrying the higher rental price.

WOOFER.

A component of many **loudspeaker** systems. It is in itself a loudspeaker, but one designed to reproduce the lower portion (**bass**) of the **audible frequency range**, from about twenty to forty **hertz**, to about four hundred to five hundred hertz. Frequencies above this range are ordinarily reproduced by loudspeakers termed **mid-range driver** and **tweeter**.

WORD. *See* **computer word**.

WORD LENGTH.

The number of either **characters** or **bits** which a **data processing** system handles as one unit or **computer word**.

WORD PROCESSOR.

A small, usually **stand-alone** device, which is composed of a **keyboard, memory**, usually a **cathode-ray tube** and, possibly, some other **peripheral equipment**, and which is capable of taking **input** and performing such simple functions as **edit**ing, **justification**, deletion, insertion, typesetting and, occasionally, translating.

WOW.

A specification appropriate to **tape recorder**s and **phonograph**s. Wow is manifested as a slow fluctuation in pitch. It is usually a form of mechanical, rather than electrical, **distortion**. Wow is caused by the same types of problems described under **flutter**.

Wow, as a specification, is often given along with flutter—e.g., "wow and flutter are...." Its citation is always a percentage of deviation from perfect speed; when wow is cited alone, it should not exceed .1 percent for a **turntable** or tape recorder.

When given as a combined wow and flutter specification, the figure should not be higher than .2 percent for a turntable, or .15 percent for a tape recorder (**open-reel**) at 7½ **inches-per-second** of **tape speed**. For an audio **cassette** device, the combined figure should not exceed .2 percent.

WRITE (COMPUTER).

Either to move **data** or **information** from **storage** to an **output** device, or to place data or information from **main storage** to **auxiliary storage**.

X. *See* **diameter**.

XEROGRAPHY. *See* **electrostatic copying**.

ZEBRA CODE. *See* **bar code**.

ZOOM LENS.

A motion picture **film** or television camera lens that can change the apparent distance between camera and subject so that although its effective **focal length** may be altered, its focus setting for the object being photographed (or televised) remains the same. It has a variable focal length that can be adjusted smoothly during a shot to create an effect somewhat like the camera moving in towards the subject from a **long shot** to **close-up** while in actuality the camera is in a fixed position.

Projectors can also use zoom lenses. Here zoom lenses are used to increase or decrease the size of the projected **image** without changing the physical distance between the projector and the **screen**.

Bibliography

GENERAL

Blatt, Joseph. "Teletext: A New Television Service for Home Information and Captioning." *Volta Review* 84, no. 4 (May 1982): 209-17.

Borrell, Jerry. "How Long the Wait until We Can Call It Television?" *Journal of Library Automation* 14, no. 1 (March 1981): 50-52.

Broderick, Richard. "Interactive Video: Why Trainers Are Tuning In." *Training* 19, no. 11 (November 1982): 46-49.

Channel 2000: Description and Findings of a Viewdata Test. Dublin, Ohio: Online Computer Library Center, Inc., 1981.

Collier, Herbert I. "Your Guide to the 1980s: Preventive Maintenance." *American School and University* 51, no. 3 (November-December 1978): 8, 11-12.

Corbin, John. *Developing Computer-Based Library Systems.* Phoenix: Oryx Press, 1981.

Criner, Kathleen, and Martha Johnson-Hall. "Videotex: Threat or Opportunity." *Special Libraries* 71, no. 9 (September 1980): 379-85.

De Gennaro, Richard. "The Computer-Based Library Systems Marketplace." *American Libraries* 9, no. 4 (April 1978): 212, 221-22.

Dymmel, Michael D. "Reacting to New Technology: The Communications Industry." *VOC ED* 57, no. 1 (January-February 1982): 41-43.

Dymmel, Michael D. "Technology in Telecommunications: Its Effect on Labor and Skills." *Monthly Labor Review* 102, no. 1 (January 1979): 13-19.

Fay, Gayle, et al. "The Electronic Schoolhouse: New Technology in the Education of the Severely Retarded." *Pointer* 26, no. 2 (Winter 1982): 10-12.

Goldstein, Charles M. "Optical Disk Technology and Information." *Science* 215, no. 4534 (February 1982): 862-68.

Heron, David W. "Waiting for the Genie: Some Requirements for Library Telefacsimile." *Journal of Academic Librarianship* 4, no. 5 (November 1978): 372-74.

Hofstetter, Fred T. "Microelectronics and Music Education." *Music Educators Journal* 65, no. 8 (April 1979): 38-45.

Kottenstette, James P., et al. *Microterminal/Microfiche System for Computer-Based Instruction: Hardware and Software Development.* Denver: Denver Research Institute of Univ. of Denver, 1980.

Lancaster, F. Wilfrid. *The Role of the Library in an Electronic Society.* Champaign: Illinois Univ. Graduate School of Library Science, 1980.

Lees, Adrian, and David Kerwin. "A Low-Cost Computerized Film Analysis System for Sports Biomechanics." *Computers and Education* 6, no. 4 (1982): 341-48.

Marcum, Deanna, and Richard Boss. "Information Technology." *Wilson Library Bulletin* 56, no. 1 (September 1981): 45-46.

McCrank, Lawrence J., ed. *Automating the Archives: Issues & Problems in Computer Applications.* White Plains, N.Y.: Knowledge Industry Publications, 1980.

McNeece, C. Aaron. "Computer-Assisted Audiovisual Training Methods for Rural Staff Development Programs." Paper presented at the National Institute on Social Work in Rural Areas, Beaufort County, S.C., July 1981.

Miller, Rosalind, and Jane Terwillegar. "Computers and School Media Services." *Clearing House* 55, no. 5 (January 1982): 210-13.

Milling, Bryan E. "Sensible Ways to Cut Your Phone Bills." *American School and University* 53, no. 9 (May 1981): 26-27.

Mischo, Lare, and Kevin Hegarty. "Videotex—The Library of the Future." *Information Technology and Libraries* 1, no. 3 (September 1982): 276-77.

Mole, Dennis. "The Videodisc as a Pilot Project of the Public Archives of Canada." *Videodisc/Videotex* 1, no. 3 (Summer 1981): 154-61.

Moore, Michael. "Educational Telephone Networks." *Teaching at a Distance*, no. 19 (Summer 1981): 24-31.

Moore, Robert C. *Home Information Systems: A Primer.* Buckhannon: West Virginia Wesleyan College, 1981.

Morrison, LaVerne, et al. *Library Learning Resources Facilities—New and Remodeled.* Austin: Texas Education Agency, 1982.

Munshi, Kiki Skagen. "Mass Media and Continuing Education: An Overview." *New Directions for Continuing Education*, no. 5 (1980): 1-14.

Neumann, Robert. "Data Banks: Opening the Door to a World of Information." *Electronic Learning* 2, no. 3 (November-December 1982): 56, 58-61, 83.

Oswitch, Pauline. *Informative Futures: Computer Simulation for Library Management*. New York: State Mutual Book and Periodical Service, 1981.

Oulton, A. J., E. Bishop, and A. J. Wood. *The Teaching of Computer Appreciation & Library Automation*. New York: State Mutual Book and Periodical Service, 1981.

Phillips, Roger. "A Public Access Videotex Library Service." *Online* 6, no. 5 (September 1982): 34-39.

Shroder, Emelie J. "Online I and R in Your Library: The State of the Art." *RQ* 21, no. 2 (Winter 1981): 128-34.

Snider, Paul B. "Great Britain's Videotex Permits TV Viewers Picture or Text Option." *Journalism Quarterly* 58, no. 2 (Summer 1981): 186-91.

Thompson, Vincent. "Videotex in Education." *Media in Education and Development* 15, no. 3 (September 1982): 118-20.

Troutner, Joanne. "Third Wave Technologies: How Do I Use Them?" *English Journal* 71, no. 3 (March 1982): 102-3.

Vaillancourt, Pauline M. *International Directory of Acronyms in Library, Information, and Computer Sciences*. New York: R. R. Bowker Co., 1980.

Wheeler, Peter. "Communication and Information Systems." *Computer Education*, no. 41 (June 1982): 17-18.

AUDIO-VISUAL

Buchanan, Gale. "AV from A to Z." *School Shop* 38, no. 5 (January 1979): 28-30, 33.

Cannon, Glenn, and Holly Jobe. *Software Maintenance*. Harrisburg, Pa.: Pennsylvania State Department of Education, Bureau of Instructional Support Services, 1978.

Cassette Sync Recorders: Parameter for Evaluation. Laboratory Test Findings. EPIE Report no. 86e. New York: Educational Products Information Exchange Institute, 1978.

Chang, Irene. "Audiotape and Videotape Basics: A Fifteen-Minute Cram Session." *NALLD Journal* 12, nos. 3-4 (Spring-Summer 1978): 39-41.

Dayton, Deane K. "A Brief History of Still Projection." *Audiovisual Instruction* 24, no. 6 (September 1979): 24-27.

Ferguson, Calum. "Audio: An Under-Rated Teaching Medium." *Media in Education and Development* 15, no. 2 (June 1982): 95-98.

Jobe, Holly, and Glenn Cannon. *A Guide to Making an Audio Tape*. Harrisburg, Pa.: State Department of Education, Bureau of Instructional Support Services, 1978.

Linder, Harris J., ed. "Instructional Media." *Journal of College Science Teaching* 8, no. 4 (March 1979): 250-54.

McAlpin, Janet. "The Overhead Projector in the Advanced Reading Class." *English Language Teaching Journal* 33, no. 3 (April 1979): 214-18.

Smith, Judson. "How to Choose Audio and Video Tape and Cassettes That Best Fit Your Needs." *Training* 15 (January 1978): 1, 43-44 et passim.

Sound Slide Projectors with Audio Cassettes. EPIE Report no. 88e. New York: Educational Products Information Exchange Institute, 1979.

Watson, Frank. "Using Audio Visual Aids: Importance of Initial Preparation." *Training Officer* 16, no. 11 (November 1980): 306-9.

Wyman, Raymond. "Safety Standards for Projectors." *Audiovisual Instruction* 24, no. 3 (March 1979): 48.

COMPUTERS

Adler, Howard. *Thirty-Four VIC-20 Computer Programs for Home, School, and Office*. Woodsboro, Md.: ARCsoft Publishers, 1983.

Aiken, Robert M. "The Golden Rule and Ten Commandments of Computer Based Education (CBE)." *Technological Horizons in Education* 8, no. 3 (March 1981): 39-42.

Allard, Kim E. "The Videodisc and Implications for Interactivity." Paper presented at the Annual Meeting of the American Educational Research Association, New York, March 1982.

Auten, Anne. "ERIC/RCS: A Guide to Purchasing a Microcomputer." *Journal of Reading* 26, no. 3 (December 1982): 267-71.

Auten, Anne. "So You Want to Buy a Microcomputer: A Guide to Purchasing." *English Journal* 71, no. 6 (October 1982): 56-57.

Banet, Bernard. "Computers and Early Learning." *Creative Computing* 4, no. 5 (September-October 1978): 90-95.

Becker, Gary. "Memorandum to Beleaguered Buyers: Topic: Microcomputer Purchases." *Electronic Education* 2, no. 2 (October 1982): 28-29.

Becker, Henry Jay. "Microcomputers: Dreams and Realities." *Curriculum Review* 21, no. 4 (October 1982): 381-85.

Bell, Frederick H. "Implementing Instructional Computing and Computer Literacy in a School or College." *AEDS Journal* 15, no. 4 (Summer 1982): 169-76.

Bell, Kathleen. "The Computer and the English Classroom." *English Journal* 69, no. 9 (December 1980): 88-90.

Bennett, Wilma E. *Checklist/Guide to Selecting a Small Computer.* New York: Pilot Industries, 1980.

Berger, Ivan. "How to Buy a Disk System." *Popular Computing* 1, no. 6 (April 1982): 92-96.

Botterell, Art. "Which Micro for Me?" *Educational Computer Magazine* 2, no. 1 (January-February 1982): 30-31, 50-51.

Brelsford, William M., and Daniel A. Relles. *Statlib: A Statistical Computing Library.* Englewood Cliffs, N.J.: Prentice-Hall, 1981.

Brown, Faith K. "A Twelve-week High School Course in Computer Use." *Science Activities* 19, no. 3 (September-October 1982): 17-20.

Buffa, Frank P., and George C. Fowler. "A Micro Computer Based Information System for Administering an Academic Program." *Computers and Education* 6, no. 4 (1982): 349-59.

Bursky, Dave. "Disk Memories: What You Should Know before You Buy Them." *Personal Computing* 5, no. 4 (April 1981): 20-27.

Buxton, Marilyn S., and Henry A. Taitt. "Microcomputers: A Creative Approach for Young Minds." In *The Computer: Extension of the Human Mind.* Proceedings of the Annual Summer Conference, Univ. of Oregon College of Education, Eugene, Oreg., July 1982.

Chandor, Anthony. *The Facts on File Dictionary of Microcomputers.* New York: Facts on File, 1981.

Clark, Ron. *Fifty-Five Color Computer Programs for the Home, School, and Office.* Woodsboro, Md.: ARCsoft Publishers, 1982.

Clark, Ron. *Fifty-Five More Color Computer Programs for the Home, School, and Office.* Woodsboro, Md.: ARCsoft Publishers, 1982.

Cole, Jim. *Fifty More Programs in BASIC for the Home, School, and Office.* Woodsboro, Md.: ARCsoft Publishers, 1981.

Cole, Jim. *Fifty Programs in BASIC for the Home, School, and Office.* 2d ed. Woodsboro, Md.: ARCsoft Publishers, 1981.

"The Computer Shopping Guide." *Instructor* 89, no. 8 (March 1980): 88-90.

Davis, Charles H., and Gerald W. Lundeen. *Illustrative Computer Programming for Libraries: Selected Examples for Information Specialists.* 2d ed. Westport, Conn.: Greenwood Press, 1981.

Dlabay, Les R. "The Educator's Guide to Computer Periodicals." *Curriculum Review* 21, no. 2 (May 1982): 144-46.

Even-Tov, Sheila. "Small Computers Take Over Giant Jobs." *American School and University* 54, no. 4 (December 1981): 3-21.

Finch, Carleton K., and David H. Dennen. *Designing a Computer Support System for School: A Handbook for Administrators.* Reading, Mass.: Addison-Wesley Publishing Co., 1982.

Flint, Glen, and Mark Dahmke. "The Atari 400." *Popular Computing* 1, no. 7 (May 1982): 74-78.

Folk, Mike. "Should Your School Get a Microcomputer?" *Mathematics Teacher* 71, no. 7 (October 1978): 608-13.

Fosdick, Howard. *Computer Basics for Librarians and Information Scientists.* Arlington: Information Resources Press, 1981.

Fosdick, Howard. *Structured PL-I Programming: For Textual and Library Processing.* Littleton, Colo.: Libraries Unlimited, 1982.

Garland, Roy, ed. *Microcomputers and Children in the Primary School.* New York: International Publications Service, 1982.

Greene, Mark M. "The Use of Micro-Computers in Educational Evaluation." In *The Computer: Extension of the Human Mind.* Proceedings of the Annual Summer Conference, Univ. of Oregon College of Education, Eugene, Oreg., July 1982.

Griffiths, Jose-Marie. *Application of Minicomputers and Microcomputers to Information Handling.* Paris: United Nations Educational, Scientific, and Cultural Organization, 1981.

Halvorsen, Nancy. "Training to Use Microcomputers for Accounting Procedures." *Journal of Business Education* 57, no. 7 (April 1982): 270-71.

Hampshire, Nick. *Library of PET Subroutines.* Rochelle Park, N.J.: Hayden Book Co., 1982.

Haney, Michael R. "The Microcomputer in the High School Science Laboratory." In *The Computer: Extension of the Human Mind.* Proceedings of the Annual Summer Conference, Univ. of Oregon College of Education, Eugene, Oreg., July 1982.

Hannaford, Alonzo, and Eydie Sloane. "Microcomputers: Powerful Learning Tools with Proper Programming." *Teaching Exceptional Children* 14, no. 2 (November 1981): 54-57.

Helms, Harry L., Jr. *Computer Language Reference Library*. Indianapolis: Howard W. Sams and Co., 1980.

Hiraki, Joan, and Oscar N. Garcia. "Setting the Stage for the Interactive Classroom of the 1980s." *Educational Computer Magazine* 1, no. 1 (May-June 1981): 20-22.

Homes, Edith. "Computers and Kids: A New Center Offers Children Hands on Experience." *Bulletin of the American Society for Information Science* 7, no. 5 (June 1981): 12-16.

Hoover, Todd, and Sandra Gould. "Computerizing the School Office: The Hidden Cost." *NASSP Bulletin* 66, no. 455 (September 1982): 87-91.

Hope, Mary. "How Can Microcomputers Help?" *Special Education: Forward Trends* 7, no. 4 (December 1980): 14-16.

House, James E. "Computers Control the Instruments of Science." *Personal Computing* 6, no. 8 (August 1982): 72-77, 80, 132.

Howell, Robert W., George J. Vensel, and Lee M. Joiner. *Educators, Parents, and Micros: How to Help Your School Get and Use Computer Power*. Kalamazoo, Mich.: Learning Publications, 1983.

Hunter, Eric J. *The ABC of BASIC: An Introduction to Programming for Librarians*. Hamden, Conn.: Shoe String Press, 1982.

Johnson, Jo Rykken. "Making the Transition to Computers Easy." *Educational Computer* 1, no. 2 (July-August 1981): 16-19.

Kock, Helen T., et al. "New Technology for Teachers." *VOC ED* 57, no. 5 (June 1982): 35-41.

Kolodziej, Leo, and John Holland. "The VIC-20." *Popular Computing* 1, no. 7 (May 1982): 96-102.

Kusack, James M., and John S. Bowers. "Public Microcomputers in Public Libraries." *Library Journal* 107, no. 20 (November 1982): 2137-41.

Lukas, Terrence. "How to Set Up an Electronic Bulletin Board." *Technological Horizons in Education* 8, no. 5 (September 1981): 50-53.

Marshall, David G. "Purchasing a Microprocessor System for Administrative Use in Schools." *AEDS Journal* 15, no. 4 (Summer 1982): 183-97.

Matthews, John I. "Considerations in Selecting Microcomputers for Instructional Design." *Journal of Industrial Teacher Education* 19, no. 1 (Fall 1981): 26-35.

McDonald, David, and Karen Gibson. "A Case Study: The Rural School District and the Microcomputer." *NASSP Bulletin* 66, no. 455 (September 1982): 75-77.

McWilliams, Peter. "The Osborne 1." *Popular Computing* 1, no. 8 (June 1982): 50-54.

Miastkowski, Stan. "Modems: Hooking Your Computer to the World." *Popular Computing* 2, no. 1 (November 1982): 88-90 et passim.

Miastkowski, Stan. "The Sinclair ZX81." *Popular Computing* 1, no. 7 (May 1982): 88-94.

Miller, Inabeth. *Microcomputers and the Media Specialist: An Annotated Bibliography.* Syracuse, N.Y.: ERIC Clearinghouse on Information Resources, 1981.

Morgan, Chris. "The Texas Instruments 99/4 Personal Computer." *on Computing* 2, no. 2 (Fall 1980): 29-30, 32-33.

Muscat, Eugene. "Microcomputers in Business Education." *Business Education World* 60, no. 4 (March-April 1980): 10-11.

Nahigian, J. Victor, and William S. Hodges. *Computer Games for Businesses, Schools, and Homes.* Boston: Little, Brown and Co., 1979.

Nicklin, R. C., and John Tashner. "Micros in the Library Media Center?" *School Media Quarterly* 9, no. 3 (Spring 1981): 168-72; 177-81.

North, Alan. *Thirty-One New Atari Computer Programs for Home, School, and Office.* Woodsboro, Md.: ARCsoft Publishers, 1982.

Okey, James R. "Computer-based Classroom Testing." *Science Activities* 19, no. 3 (September-October 1982): 29-31.

Orwig, Gary, and William Hodges. *The Computer Tutor: Learning Activities for Homes and Schools.* Cambridge, Mass.: Winthrop Publishers, 1981.

Page, Edward. *Thirty-Seven Timex 1000-Sinclair ZX-81 Computer Programs for Home, School, and Office.* Woodsboro, Md.: ARCsoft Publishers, 1982.

Pattie, Kenton, and Mary Ernst. "Finding Funds for Microcomputers." *American School and University* 54, no. 4 (December 1981): 22-25.

Patton, Robert, et al. *Computer Literacy for All High School Students.* Washington, D.C.: National Institute of Education, 1981.

Pemberton, Jeffrey K. "Should Your Next Terminal Be a Computer?" *Database* 4, no. 3 (September 1981): 4-6.

Phillips, Richard L. "Low Cost Computer Graphics in Engineering Education." *Computers and Education* 5, no. 4 (1981): 193-200.

Radin, Stephen, and Harold M. Greenberg. *Computer Literacy for School Administrators.* Lexington, Mass.: Lexington Books, 1983.

Ralston, Anthony, ed. *Encyclopedia of Computer Science.* New York: Van Nostrand Reinhold Co., 1976.

Reckord, Joshua. "Elementary Classroom Computing." In *The Computer: Extension of the Human Mind*. Proceedings of the Annual Summer Conference, Univ. of Oregon College of Education, Eugene, Oreg., July 1982.

Rorvig, Mark E. *Microcomputers and Libraries: A Guide to Technology, Products, and Applications*. White Plains, N.Y.: Knowledge Industry Publications, 1981.

Rottier, Jerry. "Reducing Computer Tension." *Instructional Innovator* 27, no. 3 (March 1982): 28.

Rowley, Jennifer. *Computers for Libraries*. Hamden, Conn.: Shoe String Press, 1980.

Saffady, William. *Computer-Output Microfilm: Its Library Applications*. Chicago: American Library Association, 1978.

Shetler, Stephen, and Pauline Shetler. "Choosing the Right Computer: A General Method." *on Computing* 2, no. 3 (Winter 1980): 67-71.

Sidman, Bernard. *Educational Computer Technology: A Manual-Guide for Effective and Efficient Utilization by School Administrators*. Palo Alto, Calif.: R and E Research Associates, 1979.

Sippl, Charles J. *Computer Dictionary*. 3d ed. Indianapolis: H. W. Sams, 1980.

Skier, Ken. "Radio Shack's Color Computer." *Popular Computing* 1, no. 7 (May 1982): 80-86.

Slezak, Tom. "Interfacing for Economical Data Gathering and Processing." *Technological Horizons in Education* 8, no. 5 (September 1981): 59-60.

Smith, Lorraine. "Choosing a Computer for Education." *Popular Computing* 1, no. 2 (December 1981): 108.

Spencer, Donald D. *BASIC: A Unit for Secondary Schools*. 2d ed. Ormond Beach, Fla.: Camelot Publishing Co., 1980.

Stevens, Dorothy Jo, and Ward Sybouts. "Computers in the Classroom." *Clearing House* 56, no. 2 (October 1982): 82-85.

Stewart, George. "How Should Schools Use Computers?" *Popular Computing* 1, no. 2 (December 1981): 104, 106, 108.

Stewart, George. "The IBM Personal Computer." *Popular Computing* 1, no. 5 (March 1982): 24-35.

Stewart, George. "The TI-99/4A." *Popular Computing* 1, no. 6 (April 1982): 57-62.

Swigger, Kathleen, and James Campbell. "The Computer Goes to Nursery School." *Educational Computer* 1, no. 2 (July-August 1981): 10-12.

Taylor, Robert, ed. *The Computer in the School: Tutor, Tutee, Tool*. New York: Columbia University Teachers College Press, 1980.

Tesch, Robert C. "Data Processing: Teaching Data Processing with a Microcomputer." *Business Education Forum* 34, no. 6 (March 1980): 20-21.

Thorkildsen, Ron. "A Microcomputer/Videodisc System for Delivering Computer Assisted Instruction to Mentally Handicapped Students." In *The Computer: Extension of the Human Mind*. Proceedings of the Annual Summer Conference, Univ. of Oregon College of Education, Eugene, Oreg., July 1982.

Tinker, Robert F. "Microcomputers in the Teaching Lab." *Physics Teacher* 19, no. 2 (February 1981): 94-105.

Tinker, Robert, and Tim Barclay. "Computers in the Science Lab." *Classroom Computer News* 2, no. 1 (September-October 1981): 22-23.

"Tracking Students with a Computer." *American School and University* 54, no. 1 (September 1981): 48.

Turner, Len. *Thirty-Six Texas Instruments TI-99-4A Programs for Home, School, and Office*. Woodsboro, Md.: ARCsoft Publishers, 1983.

Univ. of Alabama College of Education. *The Administrative Use of Microcomputers*. Technical Report. Tuscaloosa: Univ. of Alabama College of Education, 1982.

Valentine, Pamela. "Understanding Floppy Disks." *on Computing* 2, no. 1 (Summer 1980): 8-15.

Walters, Gregory M. "A Guide for the Design and Implementation of a Microcomputer-Based Learning Lab." *Educational Computer Magazine* 2, no. 1 (January-February 1982): 26-29, 47-49.

Watt, Dan. "Which Computer Should a School Buy?" *Popular Computing* 2, no. 2 (December 1982): 140-42, 144.

Welsh, Theresa. "The Radio Shack TRS-80 Pocket Computer." *Popular Computing* 1, no. 5 (March 1982): 38-42.

White, Fred. *Thirty-Three New Apple Computer Programs for Home, School, and Office*. Woodsboro, Md.: ARCsoft Publishers, 1982.

Wiggers, Thomas. "Utilization of the Apple Microcomputer for the Generation of Matching-Type Examination Questions." *Educational Computer* 2, no. 4 (July-August 1982): 32-33.

Wolverton, L. Craig. "Utilizing a Micro in the Accounting Classroom." *Business Education Forum* 37, no. 2 (November 1982): 7-8.

Woods, Rollo, and C. M. Phillips. *Managing Library Computers*. New York: State Mutual Book and Periodical Service, 1981.

MICROFORMS

Ashby, Peter. "Illustrated Reference Teaching and Beyond: The Microfiche Approach." *Microform Review* 8, no. 3 (Summer 1979): 198-201.

Burke, Pierre. "Microfilm Consolidation: The Most Economical Way to Achieve Maximum Space Utilization." *Microform Review* 11, no. 3 (Summer 1982): 176-77.

Carroll, C. Edward. *Micrographics and the Library: A Graduate Course.* Columbia: Missouri Univ. School of Library and Information Science, 1979.

Farber, Evan Ira. "The Administration and Use of Microform Serials in College Libraries." *Microform Review* 7, no. 2 (March-April 1978): 81-84.

Gunn, Michael J. "Colour Microforms and Their Application to the Visual Arts." *Microform Review* 8, no. 3 (Summer 1979): 187-92.

Laubacher, Marilyn R. *A Brief Resource Guide to Sources of Information about Microform Equipment.* Syracuse, N.Y.: ERIC Clearinghouse on Information Resources, 1979.

Raikes, Deborah A. "Microforms at Princeton." *Microform Review* 11, no. 2 (Spring 1982): 93-105.

Reibach, Lois R. *The COM Catalog: A Plan for Implementation.* Washington, D.C.: National Institute of Education, 1981.

Saffady, William. "Micrographics, Reprography, and Graphic Communications in 1979." *Library Resources and Technical Services* 24, no. 3 (Summer 1980): 283-93.

Saffady, William, and Rhoda Garoogian. "Micrographics, Reprography, and Graphic Communications in 1981. *Library Resources and Technical Services* 26, no. 3 (July-September 1982): 294-305.

Samuel, Evelyn. "Planning a Microform Center for the Art Library." *Microform Review* 8, no. 3 (Summer 1979): 160-63.

Schleifer, Harold B. "The Utilization of Microforms in Technical Services." *Microform Review* 11, no. 2 (Spring 1982): 77-92.

Sisson, Jacqueline D. "Microforms and Collection Development in Art Research Libraries." *Microform Review* 8, no. 3 (Summer 1979): 164-72.

Spaulding, Carl M. "An Update on Micrographics." *Journal of Library Automation* 11, no. 4 (December 1978): 324-28.

Veaner, Allen B. "Practical Microform Materials for Libraries: Silver, Diazo, Vesicular." *Library Resources and Technical Services* 26, no. 3 (July-September 1982): 306-8.

Zybura, Edward L. "Portable Microfiche Readers." *Bulletin of the American Society for Information Science* 7, no. 1 (October 1980): 16-18.

PHOTOGRAPHY

Busselle, Michael. *Master Photography: Take and Make Perfect Pictures.* Chicago: Rand McNally and Co., 1978.

Coleman, A. D., Patricia Grantz, and Douglas Sheer. *Photography A-V Program Directory.* Staten Island, N.Y.: Photography Media Institute, 1980.

Daniel, Robert A. "Introduction to Photography." *School Arts* 77, no. 7 (March 1978): 22-23.

Freeman, Michael. *Photo School: A Step by Step Course in Photography.* New York: American Photographic Book Publishing Co., 1982.

Galindez, Peter. "Focus on Color Photography." *Science Activities* 15, no. 1 (1978): 24-29.

Galli, Anthony John, et al. "Physics of Photography Course for Fine Arts Students: The View Camera." *American Journal of Physics* 49, no. 7 (July 1981): 641-45.

Hall, Dory. "Close the Door So the Dark Won't Leak Out." *Communication: Journalism Education Today* 14, no. 3 (Spring 1981): 12-15.

Hulme, Kenneth S. "Astronomical Photography for the Classroom." *Science Activities* 18, no. 1 (February-March 1981): 21-25.

Jones, Jim. "Photographers Should Be Aware of the Potential of the Pinhole." *Quill and Scroll* 53, no. 1 (October-November 1978): 4-7.

Logan, Larry L. *Professional Photographer's Handbook-Nikon School Edition.* Hollywood: Logan Design Group, 1981.

Lynch, Joan D., and Garth S. Jowett. *Film Education in Secondary Schools: A Study of Film Use and Teaching in Selected English and Film Courses.* New York: Garland Publishing, 1982.

Pfortner, Ray. "Getting the Most out of Your Camera ... and Your Photographs." *Nature Study* 36, nos. 1-2 (December 1982): 14-15.

Ross, Frederick C., and Rodney J. Smith. "Teaching Depth of Field Concept." *American Biology Teacher* 40, no. 1 (January 1978): 43-45.

Scheuchenzuber, H. Joseph. "Split-Image Photography for Analysis of Semisubmerged Activities." *Research Quarterly* 50, no. 1 (March 1979): 123-27.

Scudner, Peter. "Training a Green Photo Staff: Planting Ideas and Watching Them Grow." *Scholastic Editor* 59, no. 1 (September 1979): 4-7.

Seng, Mark. "Inexpensive Slide Copier." *NALLD Journal* 14, no. 1 (Fall 1979): 39-43.

Seymour, William O. "Activate the Brain: The First Step toward Quality Photographic Prints." *Quill and Scroll* 54, no. 4 (April-May 1980): 21-25.

Seymour, William. "Not for Beginning Photo Students Only: Lenses and Their Uses, Part Two." *Scholastic Editor* 59, no. 3 (January-March 1980): 16-19.

Seymour, William O. "Not for Beginning Photo Students Only: Tips on Buying a Camera, Part One." *Scholastic Editor* 59, no. 2 (November-December 1979): 20-22.

Seymour, William O. "Not for Beginning Photo Students Only: What to Look for When Buying an Electronic Flash, Part Three." *Scholastic Editor* 59, no. 4 (April-May 1980): 16-18.

Speight, Jerry. "Toy Cameras and Color Photographs." *School Arts* 79, no. 1 (September 1979): 52-54.

TELEVISION

Agostino, Don, and Jayne Zenaty. *Home VCR Owners' Use of Television and Public Television: Viewing, Recording, and Playback.* Washington: Corporation for Public Broadcasting, 1980.

Amatuzzi, Joseph R. *Television and the School.* Palo Alto, Calif.: R and E Research Associates, 1983.

Anderson, Charles. "Digital Video: How It Works, What It Can Do, and When It's Coming." *Instructional Innovator* 27, no. 3 (March 1982): 22-27.

Burke, Michael A. "Video Technology and Education: Genie in a Bottle or Pandora's Box?" *Technological Horizons in Education* 8, no. 5 (September 1981): 57-58.

Canon, Gail. "Television Captioning at the Clarke School for the Deaf." *American Annals of the Deaf* 125, no. 6 (September 1980): 643-54.

Carson, C. Herbert. "Cable Television and the Community College." *Community College Frontiers* 8, no. 1 (Fall 1979): 23-28.

Coffelt, Kenneth, and Bob Combs. *Basic Design and Utilization of Instructional Television.* 2d ed. Austin: Texas Univ. General Libraries, 1981.

Emmens, Carol A. "Large Screen Television." *School Library Journal* 28, no. 7 (March 1982): 113.

Esteves, Roberto. "Video Opens Libraries to the Deaf." *American Libraries* 13, no. 1 (January 1982): 36, 38.

Gallagher, Margaret. "The Best Use of Educational TV." *Media in Education and Development* 15, no. 3 (September 1982): 132-37.

Gothberg, Helen. *Television-Video in Libraries and Schools.* Hamden, Conn.: Shoe String Press, 1983.

Hird, John R., and Steven Balzarini. *Television Production.* New Brunswick: New Jersey Vocational-Technical Curriculum Laboratory, 1980.

Hurst, Paul. "The Utilization of Educational Broadcasts." *Educational Broadcasting International* 14, no. 3 (September 1981): 104-7.

Inouye, Ron, et al. *A Teacher's Guide to Video.* Fairbanks: Alaska Univ. Center for Cross-Cultural Studies, 1979.

Jensen, Ken, and A. J. S. Ball. "A Novelty Whose Time Has Come." *Canadian Library Journal* 38, no. 4 (August 1981): 199-205.

Kaplan, Don. "Checking out Video: The Public Library as Resource." *Media and Methods* 15, no. 9 (May-June 1979): 38-41.

Kaplan, Don. "Video: How It Works and How to Work It." *Media and Methods* 16, no. 8 (April 1980): 52-54 et passim.

Kravontka, Stanley J. "CCTV System Design for School Security." *Security World* 11, no. 2 (January 1974): 22-23, 48-49.

MacDonald, Barrie, and Kathleen M. May, eds. *Broadcasting: A Selected Bibliography.* 2d ed. London: Independent Broadcasting Authority, 1981.

Markuson, Carolyn. "Video Technology—Past and Present." *Catholic Library World* 53, no. 3 (October 1981): 112-15.

Mercer, B. Jill. *Factors Influencing the Utilization of Educational Television.* Edmonton: Alberta Educational Communications Corporation, 1980.

Middleton, John. *Cooperative School Television and Educational Change.* Washington: National Association of Educational Broadcasters, 1979.

Pulley, James B. "Classroom Ideas: Video Microprojection in Your Classroom." *Science Teacher* 46, no. 9 (December 1979): 33.

Ritter, Craig. "Two-Way Cable TV: Connecting a Community's Educational Resources." *Electronic Learning* 2, no. 2 (October 1982): 60-63.

Robinson, Sondra G. "Choosing a Color Television System." Paper presented at the annual meeting of the International Communication Association, Acapulco, May 1980.

Schipma, Peter B. "Videodisc for Storage of Text." *Videodisc/Videotex* 1, no. 3 (Summer 1981): 168-72.

Sneed, Charles. "The Videodisc Revolution: What's Ahead for Libraries?" *Wilson Library Bulletin* 55, no. 3 (November 1980): 186-89, 238.

Weggener, Joe. "A Dozen Ways to Use Video Technology in a College Education." *Educational Technology* 18, no. 1 (January 1978): 42-44.

Wood, R. Kent, and Robert D. Woolley. *An Overview of Videodisc Technology and Some Potential Applications in the Library, Information, and Instructional Sciences.* Syracuse, N.Y.: ERIC Clearinghouse on Information Resources, 1980.

WORD PROCESSING

Burrow, Elaine. *Introduction to Word Processing: Units of Instruction. Teacher's Guide.* Commerce, Tex.: Occupational Curriculum Laboratory of East Texas State Univ., 1982.

Clarke, Meredith C. "Productivity Gains from the Use of Word Processing Concepts." *Professional Engineer* 49, no. 2 (February 1979): 32-39.

Guffey, Mary Ellen, and Lawrence W. Erickson. *Business Office Practices Involving the Typewriter with Implications for Business Education Curricula.* Cincinnati: South-Western Publishing Co., 1981. Monograph.

Kock, Helen T., et al. "New Technology for Teachers." *VOC ED* 57, no. 5 (June 1982): 35-41.

McWilliams, Peter. "Word Processing." *Popular Computing* 1, no. 5 (March 1982): 52-54.

Pournelle, Jerry. "A Writer Looks at Word Processors." *on Computing* 2, no. 1 (Summer 1980): 83-87.

Press, Larry. "Word Processors: A Look at Four Popular Programs." *on Computing* 2, no. 1 (Summer 1980): 38-52.

Tilton, Rita Sloan. "The Office of the Future and Its Implications for Secretarial Training." *Journal of Business Education* 53, no. 8 (May 1978): 352-54.